SpringerBriefs in Quantitative Finance

More information about this series at http://www.springer.com/series/8784

Zorana Grbac · Wolfgang J. Runggaldier

Interest Rate Modeling: Post-Crisis Challenges and Approaches

 Springer

Zorana Grbac
Laboratoire de Probabilités et Modèles
  Aléatoires
Université Paris–Diderot (Paris 7)
Paris Cedex 13
France

Wolfgang J. Runggaldier
Department of Mathematics
University of Padova
Padova
Italy

ISSN 2192-7006   ISSN 2192-7014   (electronic)
SpringerBriefs in Quantitative Finance
ISBN 978-3-319-25383-1   ISBN 978-3-319-25385-5   (eBook)
DOI 10.1007/978-3-319-25385-5

Library of Congress Control Number: 2015955860

Mathematics Subject Classification (2010): 91G30, 91G20, 91G40, 60H30

Springer Cham Heidelberg New York Dordrecht London

Printed on acid-free paper

Springer International Publishing AG Switzerland is part of Springer Science+Business Media
(www.springer.com)

Preface

The purpose of this book is to provide a bridge between new research results motivated by the financial crisis and classical literature on interest rate modeling.

Motivation. The traditional textbooks on interest rate modeling are no longer adequate in a modern context as overviews of the techniques needed for the valuation of interest rate derivatives. In the years following the crisis, the problem of developing new models for interest rate derivatives has attracted significant attention, both from researchers working in financial institutions, as well as researchers working in academia. Various models are continually being proposed. The aim of this book is therefore two-fold. On the one hand, it aims at providing an overview of the state-of-the art techniques in modern interest rate modeling and, on the other hand, it attempts to clarify the link between these models and the classical literature. From the practical point of view, the importance of up-to-date interest rate models can be best illustrated when viewing the fixed-income market as a part of the global derivatives market. According to the yearly statistics provided by the Bank for International Settlements, the notional amounts outstanding each year for over-the-counter (OTC) interest rate derivatives sum up to 80 % of the total trade volume in OTC derivatives ($505 trillion out of the total volume of $630 trillion corresponded to interest rate derivatives in 2014).

Audience. The book is intended to serve as a guide for graduate students, researchers, and practitioners interested in the paradigm change that affected all fixed-income markets due to the financial crisis. More generally, we intend to address people who are already quite knowledgeable of mathematical finance, who have some familiarity with classical interest rate theory, but who have little or no familiarity with issues on multi-curve modeling. As mathematical prerequisites we expect the reader to have a basic knowledge of probability theory together with notions from stochastic processes and stochastic calculus that are commonly used in the mathematical finance literature.

Approach. Our approach to multiple curves consists mainly in modeling for the purpose of pricing interest derivatives, rather than in view of hedging and/or risk management. It corresponds to what in the classical single-curve interest rate theory is called "martingale modeling" in the sense that the models are defined under a martingale measure Q that has as numéraire the money market account and has to be calibrated to each specific basket of products, given that the market as such is incomplete. From this measure Q, and using the discount curve, one can then derive the various forward measures used for pricing of interest rate derivatives. After calibration, this will lead to unique prices. We can and shall follow the same procedure also in the multi-curve setup after justifying (see Sect. 1.3.1 below) the choice of a single specific curve for discounting future cash flows and with it the money market account and thus also a reference martingale measure Q.

For simplicity of exposition we limit ourselves to Wiener-driven models, but extensions to jump-diffusions can in most cases be obtained in a relatively straightforward manner. In each of the chapters, when it comes to pricing, we consider only what according to some of the literature is called "clean valuation" and in Sect. 1.2.3 of the introductory Chap. 1 we justify this choice. Even limiting ourselves to clean valuation, there are various possible approaches that one can find in the literature over the past years. Multi-curve modeling is very recent as a research topic and it is still early to evaluate the advantages of one multi-curve approach with respect to others; we therefore opted to limit ourselves to an overview.

Structure. In the classical pre-crisis interest rate theory one considers various models and model classes and there does not exist a single model that is uniformly better than the others. Different model classes are in fact suited for different situations and products. Consequently, also when passing to multi-curve modeling, one considers various model classes as well. The following major classical interest rate model classes have so far found an extension to the multi-curve setting and form the basis for this monograph: (i) short-rate and rational pricing kernel models; (ii) forward rate models (Heath–Jarrow–Morton setup); (iii) Libor market models (more generally, market forward rate models). In the pre-crisis setting there are also other interest rate models that are briefly cited in Chap. 1. The above ordering of the three classes reflects a "bottom-up" point of view and in our exposition we follow this ordering. In fact, starting from the introductory Chap. 1 where we explain the main notions and concepts related to the post-crisis fixed-income markets, we then proceed with three chapters as follows:

Chapter 2: This chapter concerns mainly the classical, strict-sense short-rate models, but also some wider sense short-rate models represented by the rational pricing kernel models. For the strict-sense short-rate models, we consider for each tenor a short-rate spread to be added to the short rate from the outset. For the dynamics of the short rate and the spreads we consider factor models that belong to the exponentially affine or the exponentially quadratic model classes. We develop in detail the results for the exponentially affine class, for which we are able to obtain closed or semi-closed formulas for the prices of linear and optional interest rate derivatives. For linear derivatives we are able to compare directly single curve (pre-crisis) and multiple curve (post-crisis) derivative values by means of an adjustment factor. Finally, we summarize recent multiple curve extensions of rational pricing kernel models.

Chapter 3: This chapter concerns forward rate models in a Heath–Jarrow–Morton (HJM) setup. Similarly to the short rate and its additive spreads in Chap. 2, here we consider the reference forward OIS rate and the corresponding spreads. Major emphasis is put on obtaining arbitrage-free models by deriving for them no-arbitrage conditions in the form of a "drift condition" analogous to the classical HJM condition. Since the HJM framework is situated in between the short-rate models and the Libor market models (LMMs), we proceed essentially along two ways: (i) mimicking the LMMs by a hybrid LMM–HJM approach, where we consider a linear transform of the Libor rate that is modeled directly under the standard martingale measure, but by its definition has to be a martingale under the forward measure; this then leads to no-arbitrage conditions. We shall call these "real no-arbitrage conditions" in the sense that they represent intrinsic no-arbitrage conditions in relation to the basic traded assets that are FRA contracts; (ii) the other alternative consists in reproducing the pre-crisis relationship between discretely compounding forward rates and bond prices also for the forward Libor rates in the post-crisis setup, but replacing the standard zero coupon bonds by fictitious ones that are supposed to be affected by the same risk factors as the Libor rates. In this latter context we obtain no-arbitrage conditions analogously to point (i) by imposing that the ratio of fictitious bond prices in this relationship has to be a martingale under the forward measure. In addition to this, we also discuss "pseudo no-arbitrage conditions" by assigning different interpretations to the fictitious bond prices, in particular via a credit risk and a foreign exchange analogy. The last part of the chapter concerns interest rate derivative pricing in this HJM context.

Chapter 4: This chapter presents basically an overview of two major existing approaches to obtain multi-curve models on a discrete tenor in the spirit of the Libor market models. The first approach concerns a series of papers by Mercurio and co-authors, as well as by authors related to them, in which multiple curve extensions of the classical LMMs are developed. The other approach is concerned with an affine Libor model for multiple curves. The spreads in the above approaches are additive; we conclude the chapter by mentioning approaches based on modeling multiplicative spreads.

The material presented in these chapters corresponds to a selection that we had to make among the possible material to fit into the allowable size of the monograph within the "SpringerBriefs" series. Among the topics that we were not able to include we would like to mention the issues of numerical implementation and calibration of the presented models, for which we shall always refer to corresponding articles dealing with this, as well as the questions of hedging and risk management in the multi-curve environment, which have currently been less well studied in a systematic way in the available literature, although they are of utmost practical importance. In particular, in view of hedging, one might also raise the issue of defining the price of a derivative as a cost of its hedging portfolio and a related issue of possible non-uniqueness of prices. As mentioned above, here we opt rather for the approach based on martingale modeling, where an existence of a martingale measure is assumed, the models are developed under this measure, and the prices defined as corresponding conditional expectations. This approach has the advantage of allowing to compute various post-crisis valuation adjustments such as CVA, which have to be computed for the whole aggregate portfolio of derivatives between two counterparties, and thus require a unified pricing method for all derivatives.

During the preparation of the manuscript, in the European markets we have witnessed a continuous important decrease in the level of all interest rates, as well as the appearance of negative interest rates, firstly only for the Swiss Franc, but more recently also for the Euro. This phenomenon has been observed for several months already and it has by now occurred not only at the shortest end of the interest rate curve, but also in the midterm rates. Negative rates arise because of frictions not addressed by the models, such as the "cost of carry" associated with keeping large amounts of cash. Due to this current market situation we are prompted to slowly readjust one of the long-standing modeling axioms that the interest rates should be positive. This is an interesting modeling situation, where the models in which the interest rates can become negative suddenly seem to be perfectly suited for the task at hand. One should still be cautious when addressing this issue because, even though negative rates have been observed, the multiple curve spreads still remain positive, so ideally one would need a model which combines both of these features. In this book we present some models that ensure positive interest rates and some models that do not have this property and, as mentioned above, we do not evaluate the approaches based on this quality. However, we do mention when discussing certain models providing positive interest rates that they can be modified without increasing the level of their complexity to allow for the rates to fall below zero. It is still left to be seen if the negative rate phenomenon will persist in the future as well and will become a standing modeling requirement, such as it was the case with the multiple interest rate curves.

Literature. The literature on interest rates is too vast to mention it all. Our monograph concerns multiple curves. While we made an effort to cite the relevant literature for interest rates in general, we found it most natural to concentrate mainly on the literature that concerns models that so far have found an extension to the multi-curve setting. We have tried to be as complete as possible and we apologize for having possibly overlooked some relevant literature. It is also not an easy task to keep track of all work, especially the more recent, since the subject is currently in rapid evolution and some of the key work after the crisis has been made inside the investment banks, and cannot be accessed as long as it remains inside the companies. Furthermore, it would bring us too far if for each concept we would trace back its evolution over time in the literature and thus we limited ourselves to references that in some sense summarize previous achievements. On the other hand, given that one of our purposes is to provide an overview of the state-of-the-art of multi-curve modeling, instead of taking the approach of addressing "who did what," it was natural to make repeated references since each reference does not treat only a single topic, but touches upon various arguments.

Acknowledgments. First of all we would like to thank Matheus Grasselli for having invited us to write this monograph. In fact, after having received the invitation by Matheus, Wolfgang Runggaldier had personally suggested to invite Zorana Grbac to join him as a co-author. Her gratitude extends thus to both Matheus and Wolfgang for the joint invitation.

We would also like to thank the entire Editorial Board of "SpringerBriefs in Quantitative Finance" and the Springer staff, in particular Ute McCrory, for having supported us, as well as the anonymous referees, appointed by Springer, for their valuable feedback. We would furthermore like to thank various colleagues who provided us with very useful comments and remarks on the entire or also only parts of the monograph, among them Laurence Carassus, Stéphane Crépey, Nicole El Karoui, Ernst Eberlein, Claudio Fontana, Fabio Mercurio, Antonis Papapantoleon, Mathieu Rosenbaum, Martin Schweizer, and Peter Tankov. Finally, we would also like to acknowledge financial support for our meetings coming from our research units at the Laboratoire de Probabilités et Modèles Aléatoires, Université Paris Diderot and the Department of Mathematics of the Università di Padova, as well as from the research project "Post-Crisis Models for Interest Rate Markets" financed by the Europlace Institute of Finance.

Paris, France   Zorana Grbac
Padova, Italy   Wolfgang J. Runggaldier
September 2015

Contents

Chapter 1
Post-Crisis Fixed-Income Markets

The terminology *fixed-income market* designates a sector of the global financial market on which various interest rate-sensitive instruments, such as bonds, forward rate agreements, swaps, swaptions, caps/floors are traded. *Zero coupon bonds* are the simplest fixed-income products, which deliver a constant payment (often set to one unit of cash for simplicity) at a pre-specified future time called maturity. However, their value at any time before maturity depends on the stochastic fluctuation of interest rates. The same is true for other fixed-income derivatives. Fixed-income instruments represent the largest portion of the global financial market, even larger than equities. Developing realistic and analytically tractable models for the dynamics of the term structure of interest rates is thus of utmost importance for the financial industry. From the theoretical point of view, interest rate modeling presents a mathematically challenging task, in particular due to the high-dimensionality (possibly even infinite) of the modeling objects. In this sense interest rate models substantially differ from equity price models.

The credit crisis in 2007–2008 and the Eurozone sovereign debt crisis in 2009–2012 have impacted all financial markets and have irreversibly changed the way they functioned in practice, as well as the way in which their theoretical models were developed. One may thus distinguish between a pre-crisis and a post-crisis setting. The key features that were put forward by the crises are *counterparty risk*, i.e. the risk of a counterparty failing to fulfill its obligations in a financial contract, and *liquidity* or *funding risk*, i.e. the risk of excessive costs of funding a position in a financial contract due to the lack of liquidity in the market. The fixed-income market has been particularly concerned by both of these issues. The reason for this is the following: the underlying interest rates for most fixed-income instruments are the market rates such as Libor or Euribor rates and the manner in which the market quotes for these rates are constructed (see Sect. 1.1 for details) reflects both the counterparty and the liquidity risk of the interbank market. Inspection of quoted prices for related instruments reveals that the relationships between Libor rates of different maturities

© The Author(s) 2015
Z. Grbac and W.J. Runggaldier, *Interest Rate Modeling: Post-Crisis Challenges and Approaches*, SpringerBriefs in Quantitative Finance, DOI 10.1007/978-3-319-25385-5_1

that were previously considered standard, and held reasonably well before the crisis, have broken down and nowadays each of these rates has to be modeled as a separate object. Moreover, significant spreads are also observed between Libor/Euribor rates and the swap rates based on the so-called overnight indexed swaps (OIS), which were following each other very closely before the crisis. Simultaneous presence of these various interest rate curves is referred to in the current literature as the *multiple curve* issue and the post-crisis interest rate models are often referred to as the *multiple curve models* (multi-curve models). This book aims at providing a guide for development of interest rate models in line with these changes, accompanied by a detailed overview of some current research articles, in which, to the best of our knowledge, such a development has been studied. The recent monograph by Henrard (2014) and the article collection Bianchetti and Morini (2013) also concern post-crisis multiple curve modeling, reflecting in particular the practitioners' perspective.

Pre-crisis interest rate models can be divided into various classes, in particular the following: the short-rate models, where the short interest rate is modeled, including pricing kernel models; the Heath–Jarrow–Morton (HJM) framework, where the zero coupon bond prices, or equivalently, the forward instantaneous continuously compounded rates are modeled; the Libor market models, where the market forward rates are modeled directly. The books by Björk (2009), Brigo and Mercurio (2006), Cairns (2004), Filipović (2009), Hunt and Kennedy (2004) and Musiela and Rutkowski (2005) provide an excellent introduction to interest rate theory, as well as an extensive overview of the existing modeling approaches in this field. The first short-rate models were introduced in the seminal papers by Vasiček (1977), Cox et al. (1985) and Hull and White (1990). The HJM framework was developed in Heath et al. (1992). Furthermore, rational pricing models were pioneered by Flesaker and Hughston (1996), models using potential approach were developed in Rogers (1997) and Markov functional models in Hunt et al. (2000). Finally, Libor market models were proposed by Brace et al. (1997) and Miltersen et al. (1997) and later developed further especially by practitioners. Various extensions and generalizations of all these model classes can be found in the literature, which is too vast to mention it all here.

In order to give a first flavor and illustrate the issues presented above, we shall give a closer look at a prototypic interest rate derivative, a forward rate agreement, which is a building block for all linear interest rate derivatives and is also related to the underlying rate of nonlinear derivatives, as its price represents the market expectation about the future value of the Libor rate. Precise definitions of all the notions used below, as well as a more detailed treatment of the pricing of FRAs will be presented in Sect. 1.4.1.

Let us firstly recall the classical, pre-crisis connections between zero coupon bonds, FRA rates and Libor rates. As mentioned above, a zero coupon bond is a financial contract which delivers one unit of cash at its maturity date $T > 0$. Its price at time $t \leq T$, denoted by $p(t, T)$, represents therefore the expectation of the market concerning the future value of money. Obviously, $p(T, T) = 1$. Traditionally, interest rates are defined to be consistent with the zero coupon bond prices $p(t, T)$. For discretely compounding forward rates this leads to

$$F(t; T, S) = \frac{1}{S-T} \left(\frac{p(t, T)}{p(t, S)} - 1 \right), \quad t < T < S \qquad (1.1)$$

This formula can also be justified as representing the *fair fixed rate* at time t of a *forward rate agreement* (FRA), where the floating rate received at time S is

$$F(T; T, S) = \frac{1}{S-T} \left(\frac{1}{p(T, S)} - 1 \right) \qquad (1.2)$$

The rate (1.1) is therefore also called the *FRA rate*. Note that we have assumed, without loss of generality, the notional equal to 1 here, as we are interested only in the rates. The arbitrage-free price at time t of such an FRA is, using the forward martingale measure Q^S (see Sect. 1.3.2),

$$P^{FRA}(t; T, S, R) = p(t, S)(S - T) E^{Q^S} \{F(T; T, S) - R \mid \mathscr{F}_t\} \qquad (1.3)$$

where R denotes the fixed rate of the FRA. This price is zero for

$$R = E^{Q^S} \{(F(T; T, S) \mid \mathscr{F}_t\}$$

$$= E^{Q^S} \left\{ \tfrac{1}{S-T} \left(\tfrac{p(T,T)}{p(T,S)} - 1 \right) \Big| \mathscr{F}_t \right\} = \tfrac{1}{S-T} \left(\tfrac{p(t,T)}{p(t,S)} - 1 \right)$$

Therefore, one obtains the following key relation between the bond prices and the forward rates thanks to the connection between the floating rate and the zero coupon bond prices (1.2):

Definition 1.1 The *discretely compounded forward rate* at time $t \geq 0$ for the future time interval $[T, S]$, where $t \leq T \leq S$, is the rate given by

$$F(t; T, S) = E^{Q^S} \{(F(T; T, S) \mid \mathscr{F}_t\} = \frac{1}{S-T} \left(\frac{p(t, T)}{p(t, S)} - 1 \right) \qquad (1.4)$$

Before the crisis, the Libor rate, which is an interest rate obtained as an arithmetic average of submitted daily quotes from a panel of banks participating in the London interbank market (see Sect. 1.1), was assumed to be equal to the floating rate defined using zero coupon bond prices, i.e.

$$L(T; T, S) = F(T; T, S) = \frac{1}{S-T} \left(\frac{1}{p(T, S)} - 1 \right) \qquad (1.5)$$

where $L(T; T, S)$ stands for the Libor rate at time T for the period $[T, S]$. This was rightfully done so, since the Libor panel, which is refreshed on a regular basis in such a way that the banks of deteriorating credit quality are replaced with the banks of a better credit quality, contained virtually no counterparty and liquidity risk, making thus plausible the assumption of risk-freeness, which is implicitly made when assuming (1.5). Consequently, the FRA rate $F(t; T, S)$ was also called the *forward Libor rate* (since it represented the market expectation of the future value of the Libor rate) and often denoted by $L(t; T, S)$. Hence, the forward Libor rate was given either as a conditional expectation of the spot Libor rate under the forward martingale measure, or expressed using the quotient of the bond prices, cf. (1.4):

$$L(t; T, S) = E^{Q^S} \{(L(T; T, S) \mid \mathscr{F}_t\} = \frac{1}{S - T} \left(\frac{p(t, T)}{p(t, S)} - 1 \right) = F(t; T, S) \quad (1.6)$$

Due to the crisis and in view of the manner in which the spot Libor rate quotes are produced, in the post-crisis market the assumption that the Libor rate is free of various interbank risks is no longer sustainable and hence the connection (1.5) to the zero coupon bonds, which are assumed risk-free, is lost:

$$L(T; T, S) \neq \frac{1}{S - T} \left(\frac{1}{p(T, S)} - 1 \right)$$

The question what these zero coupon bonds are in the post-crisis setup is also far from a trivial one and is tackled in Sect. 1.4.4; here for simplicity we do not enter into it. Let us now consider again the same type of the forward rate agreement as above, to exchange a payment based on a fixed interest rate R against the one based on the spot Libor rate $L(T; T, S)$, cf. Definition 1.3. The payoff of the FRA at maturity S being equal to

$$P^{FRA}(S; T, S, R) = (S - T)(L(T; T, S) - R)$$

its value at time $t \leq T$ is calculated as the conditional expectation with respect to the forward measure Q^S associated with zero coupon bond $p(\cdot, S)$ as numéraire and is given by

$$P^{FRA}(t; T, S, R) = p(t, S)(S - T)E^{Q^S} \{L(T; T, S) - R|\mathscr{F}_t\}$$

We use the same symbol for the value of the FRA here as in (1.3) since it is basically still the same contract, only the underlying rate $L(T; T, S)$ does not satisfy (1.5) anymore. Hence, the key quantity is the conditional expectation of the spot Libor rate that we denote by $L(t; T, S)$ and define

Definition 1.2 The *forward Libor rate* at time $t \geq 0$ for the future time interval $[T, S]$, where $t \leq T \leq S$, is the rate given by

$$L(t; T, S) := E^{Q^S} \{L(T; T, S) \mid \mathscr{F}_t\}, \qquad 0 \leq t \leq T \leq S$$

The crucial difference with respect to the classical pre-crisis forward Libor rate is the following one:

$$L(t; T, S) = E^{Q^S} \left[(L(T; T, S) \mid \mathscr{F}_t \right] \neq \frac{1}{S-T} \left(\frac{p(t,T)}{p(t,S)} - 1 \right) = F(t; T, S) \quad (1.7)$$

Typically, the former quantity will be greater than the latter and this provides the very first example of the post-crisis spreads (or equivalently multiple curves):

$$S(t; T, S) := L(t; T, S) - F(t; T, S) \qquad (1.8)$$

This spread depends, moreover, also on the difference $\Delta := S - T$, i.e. the length of the period to which the Libor rate applies, also known as *tenor*. In practice, the tenor Δ ranges from one day to several months (up to twelve months) and the observed market spreads are typically increasing with respect to the tenor, i.e. the function $\Delta \mapsto S(t; T, T + \Delta)$ is an increasing function for fixed t and T. This and other types of spreads will be defined in Sect. 1.4.4.

1.1   Types of Interest Rates and Market Conventions

In this section we describe the most important market rates and their main characteristics. Note that all these rates are quoted as *annualized* rates. This means for example that a quote of 1 % for a 3-month interest rate corresponds to the following interest: 1 unit of cash invested at this rate yields $1 + \frac{3}{12}0.01 = 1.0025$ units of cash in 3 months.

1.1.1   Basic Interest Rates: Libor/Euribor, Eonia/FF and OIS Rates

The most widely known market rates are the Libor rates because they are reference rates for a variety of fixed income products, but even a person without any experience in the financial industry might have seen this rate as an underlying floating rate in bank loans for example. Therefore, we begin this section by giving a description and the main characteristics of the Libor rates. The LIBOR stands for London Interbank Offered Rate and the description below is taken from the ICE Benchmark Administration (IBA) webpage https://www.theice.com/iba/libor, which is an independent entity administering the Libor as of February 1st, 2014. The Financial Conduct Authority (FCA) serves as a regulator, which supervises the panel banks and has a power to take individuals to court for benchmark-related misconduct. We quote from the ICE Benchmark Administration:

"ICE Libor is designed to reflect the short term funding costs of major banks active in London, the world's most important wholesale financial market. Like many other financial benchmarks, ICE Libor (formerly known as BBA Libor) is a *polled* rate. This means that a panel of representative banks submits rates which are then combined to give the ICE Libor rate. Panel banks are required to submit a rate in answer to the ICE Libor question: *At what rate could you borrow funds, were you to do so by asking for and then accepting inter-bank offers in a reasonable market size just prior to 11 a.m.?* Although banks now use transaction data to anchor their submissions, having a polled rate is crucial to ensure the continuous publication of such a systemic benchmark, even in times when liquidity is low and there are few transactions on which to base the rate. Currently only banks with a significant London presence are on the ICE Libor panels, yet transactions with other non-bank financial institutions can often inform panel banks' submissions. *Reasonable market size* is intentionally unquantified. The definition of an appropriate market size depends on the currency and tenor in question, as well as supply and demand. The current wording therefore avoids the need for frequent and confusing adjustments. 11 a.m. was chosen because it falls in the most active part of the London business day. It is also sufficiently early in the day to allow the users of ICE Libor to use each day's rates for valuation processes, which may take place in the afternoon. All ICE Libor rates are quoted as annualized interest rates. This is a market convention. For example, if an overnight Pound Sterling rate from a contributor bank is given as 0.5000 %, this does not indicate that a contributing bank would expect to pay 0.5 % interest on the value of an overnight loan. Instead, it means that it would expect to pay 0.5 % divided by 365."

Note that the ICE Libor rates are produced each business day for five different *currencies* (US Dollar, Euro, British Pound Sterling, Japanese Yen, Swiss Franc) and seven *maturities* (1 day, 1 week, 1, 2, 3, 6 and 12 months). For each currency, a different panel of representative banks is selected, ranging from 11 to 18 banks. The ICE Libor daily quote for each currency and each maturity is the "trimmed arithmetic mean" of all of the panel banks' submissions, i.e. the highest and lowest 25 % of the submissions are removed and the rest is averaged.

In the Eurozone, an interest rate with very similar features to those of the Libor is called the *Euribor*, see http://www.emmi-benchmarks.eu for details. The entity administering the Euribor is the European Money Markets Institute (EMMI) as of June 20th, 2014. Similarly to the Libor, the Euribor is also produced from quotes submitted by a panel of banks from EU countries, as well as large international banks from non EU-countries, participating in Eurozone financial operations. The choice of banks quoting for Euribor is based on market criteria and the panel consists currently of 26 contributing banks, which submit a rate at which they believe "*Euro interbank term deposits are offered by one prime bank to another prime bank within the EMU zone*". The Euribor rates are quoted for eight different *maturities* (1 and 2 weeks, 1, 2, 3, 6, 9, 12 months) and are "*calculated at 11:00 a.m. (CET) for spot value (T+2)*" as a trimmed average of the quotes submitted by the panel banks. Figure 1.1 displays the historical series of the Euribor rates for maturities 1, 2 and 3 months. The starting month in the graph is January 2010 and the last month is September 2015.[1]

---

[1] The figure is taken from http://www.emmi-benchmarks.eu/euribor-org/about-euribor.html.

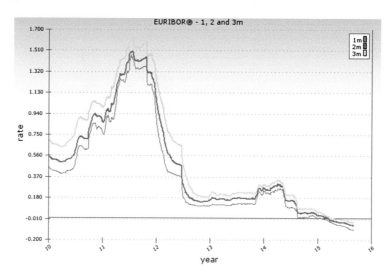

Fig. 1.1   Historical series of the Euribor rates for maturities 1, 2 and 3 months. The starting month in the graph is January 2010 and the last month is September 2015

The reference rate for the shortest maturity of one day in the Eurozone is the Eonia (Euro OverNight Index Average) rate. The Eonia rate is computed as a *"weighted average of all overnight unsecured lending transactions in the interbank market, undertaken in the European Union and European Free Trade Association (EFTA) countries by the panel banks. It is reported on an act/360 day count convention and is displayed to three decimal places"*. The panel banks for the Eonia rate are the same as the ones for the Euribor. Note that these banks contribute daily data on their total volume of transactions in unsecured overnight money and the average interest rate for this daily volume; the Eonia rate is then calculated from these contributions as a weighted average interest rate, where the weighting is done according to the transaction volumes of the contributors. In the left graph of Fig. 1.2 the yearly average values of the Euribor rates with maturities 1, 2 and 3 months are displayed and in the right graph the average Eonia rates.[2]

A corresponding overnight rate in the United States to the Eonia rate in the Eurozone is the *Federal Funds* (FF) effective rate, which is the weighted average across all overnight transactions between depository institutions trading balances held at the Federal Reserve, which are called federal funds. Similarly to the transactions contributing to the Eonia rate, these transactions are also unsecured.

Finally, the name *OIS rate* refers to a market swap rate of an *overnight indexed swap* (OIS), which is, as any interest rate swap, defined on a discrete tenor structure and in which, at every tenor date, payments based on a fixed rate are exchanged for payments based on a floating rate. This floating rate is a discretely compounded

---

[2]The figures are taken from http://www.emmi-benchmarks.eu/euribor-org/about-euribor.html and http://www.emmi-benchmarks.eu/euribor-eonia-org/about-eonia.html.

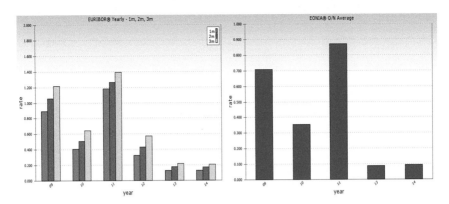

Fig. 1.2 *Left* Yearly average for Euribor rates with maturities 1, 2 and 3 months, 2009–2014. *Right* Yearly average for Eonia rates, 2009–2014

rate obtained by compounding the overnight rates over the corresponding intervals between two subsequent tenor dates (the reason why the swap is called overnight indexed swap). The reference overnight rate in the Eurozone is the Eonia and in the US the FF rate. Overnight indexed swaps and corresponding OIS rates are precisely defined and studied in detail in Sect. 1.4.4.

1.2   Implications of the Crisis

As already mentioned in the introduction to this chapter, a number of anomalies that were not previously observed in the fixed income markets appeared due to the financial crisis. The interest rates whose dynamics were very closely following each other have started to diverge substantially, thus prompting the introduction of various spreads measuring this divergence.

1.2.1   Spreads and Their Interpretation: Credit and Liquidity Risk

The first type of post-crisis spreads, mentioned already at the beginning of the chapter, concerns the spreads between the Libor rates and the OIS rates of the same maturity (see Sect. 1.4.4 and in particular, Eq. (1.33) for the precise definition), which have been far from negligible since the crisis. Moreover, also the spreads between the swap rates of the Libor-indexed interest rate swaps and the OIS rates (see Sect. 1.4.4, Eq. (1.37) for definition) have appeared. The former type of spreads is known as the *Libor-OIS spread* and the latter as the *Libor-OIS swap spread*. In Fig. 1.3[3] (left)

---

[3]The figure is taken from Crépey et al. (2012).

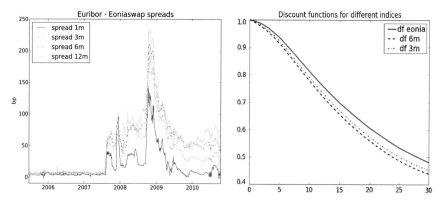

Fig. 1.3 *Left* Historical Euribor-Eonia swap spreads 2005–2010. *Right* Discount curves bootstrapped on September 2, 2010

the historical Euribor-Eonia swap spreads in the period 2005–2010 are plotted for maturities ranging from 1 month to 12 months. As one can clearly see, before the crisis these spreads were practically negligible, whereas at the peak of the crisis they were greater than 200 basis points for some maturities. Furthermore, the Libor rates of different maturities have exhibited notably diverse behavior, reflected in *basis swap spreads*, which are appearing in connection to basis swaps (cf. Sect. 1.4.5 for definition of a basis swap and Eq. (1.40) for the related basis swap spread). Figure 1.4[4] shows the evolution of the Euribor-OIS spread, as well as the 3-month versus 6-month basis swap spread, for a time period from January 2004 until April 2014. As the graphs indicate, even after 2012 these spreads did not revert back to their pre-crisis levels. This is why practitioners nowadays tend to produce different yield curves for different tenors; compare Remark 1.2 and see Fig. 1.3 (right), which displays discount curves related to the Eonia swap rates, the 3-month and the 6-month Euribor rates.

The question that immediately comes to mind is if such observed large spreads present arbitrage opportunities in the market. However, as pointed out by Chang and Schlögl (2015), these spreads have persisted since the crisis, implying that such opportunities have not been exploited. The reason is that the spreads are due to the various risk levels, therefore Chang and Schlögl (2015) conclude that the expected gains from a possible arbitrage have to be offset by the expected losses at the different risk levels, which would imply that those arbitrage opportunities are only illusory and hence cannot be exploited.

In the chapters below we shall present models for the dynamic evolution of the spreads. Notice that the first approaches to spread modeling considered them as deterministic quantities (see Henrard 2007, 2010). The advantage of this deterministic modeling is that it allows the pre-crisis single curve pricing formulas for both linear and optional derivatives to be applied in the same form also to the multiple curve

---

[4]The figure is taken from Grbac et al. (2014).

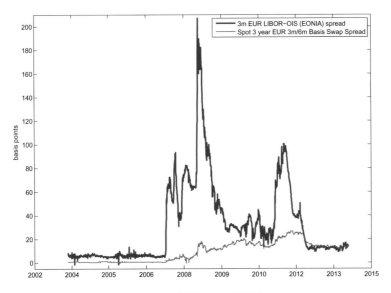

Fig. 1.4 Spread Development from January 2004 to April 2014

setup. In particular, for the case of optional derivatives, it suffices to modify the strike by simple deterministic shifting or scaling (see also more detailed comments in Sect. 4.3). For this reason, this way of modeling seems to be still often used in practice in spite of contrary evidence from empirical data (see Fig. 1.4), as well as of possible anomalies introduced by these simplistic assumptions. One such anomaly is mentioned e.g. in Mercurio and Xie (2012) (see also the last part of Sect. 4.1.1) who point out that a deterministic spread assumption would wrongly lead to a zero price for out-of-the-money Libor-OIS swaptions.

The various risks driving the spreads are the risks related to the interbank market, in particular to the banks participating in the Libor panel. Therefore, these risks are sometimes jointly referred to as *interbank risk*. One important component of this risk is default risk. The rolling construction of the Libor panel is intended to reduce the possibility of actual defaults within the panel. However, the deterioration of the credit quality of the Libor contributors during the length of a Libor-based loan is greater with longer tenors. Moreover, interbank risk arises also from liquidity risk. Strategic gaming can also play a role (Michaud and Upper 2008). Such considerations might from time to time incite a bank to declare as its Libor contribution a number different from its internal conviction regarding "The rate at which an individual Contributor Panel bank could borrow funds, were it to do so by asking for and then accepting interbank offers in reasonable market size, just prior to 11.00 London time" (the definition of Libor). All this results in a spread between the Libor rates of different tenors (OIS rates in the limiting case of an overnight tenor).

Filipović and Trolle (2013) analyze the decomposition of the interbank risk driving spreads into default and liquidity risk components. Using a data set covering the period August 2007–January 2011, the authors show that the default component is overall the main dominant driver of interbank risk, except for short-term contracts in the first half of the sample (see Figs. 3 and 4 in their paper). The remaining risk is attributed to liquidity risk, an observation made in Morini (2009) as well. The liquidity risk component driving the Libor-OIS spreads is studied and explicitly modeled in Crépey and Douady (2013). A recent work by Gallitschke et al. (2014) provides an endogenous explanation of spreads and constructs a structural model for Libor rates, deriving them from fundamental risk factors, namely: interest rate risk, credit risk and liquidity risk. The emphasis is on liquidity risk, which is shown to be mainly induced by the tenor basis.

1.2.2 From Unsecured to Secured Transactions

Broadly speaking, in finance the term *collateral* refers to assets or cash posted by a borrower to a lender in order to secure a loan. In other words, if for various reasons the borrower fails to make the promised loan payments, the lender can cover (partially or fully) the occurred losses using the collateral. If this has not been the case, the collateral is returned to the borrower after the loan has been fully repaid, together with the possibly accumulated interest. Collateral can be posted also in connection with various other financial transactions which expose one or both counterparties to the risk of non-payment (default risk). Since collateral thus provides certain security, a financial transaction with posted collateral is called *secured*, as opposed to an *unsecured* transaction which does not have collateral. The question of unsecured versus secured transactions in financial markets gained extreme importance in view of the previously discussed counterparty risk. Since it is by now generally understood that no financial institution is "too big to default", various mechanisms of reducing the exposure to counterparty risk in OTC derivatives have been put forward.

In particular, a large number of OTC bilateral contracts is *collateralized*, i.e. the value of the contract is periodically marked-to-market and the party whose position has lost in value has to post collateral in the collateral account. The posted collateral remains the property of the collateral payer and is remunerated. In case no default occurs during the lifetime of the contract, the collateral provider receives it back, together with the accumulated interest. The details regarding the frequency of posting the collateral, eligible currencies and securities which can serve as collateral, specifications of close-out cash flows are all described in CSA (Credit Support Annex) agreements. The ISDA (International Swaps and Derivatives Association) nowadays provides standard CSAs for OTC derivatives, which "is part of ISDA's continuing efforts to increase efficiency and improve standardization in the OTC derivatives markets". These standard CSAs in particular promote adoption of the OIS discounting.

An alternative to collateral posting directly between two counterparties is "central clearing" of a bilateral contract, which is done via *central counterparties* (CCPs) or *clearing houses*. Nowadays many financial contracts are cleared in this way and a number of CCPs exist specializing in particular types of markets and products. For example, some of the main CCPs for interest rate swaps and credit default swaps are SwapClear, ICE Trust US and ICE Clear Europe. A short description of functioning of a CCP given here is based on the article by Heller and Vause (2011).

The role of a CCP is to act as an intermediary between two counterparties in a bilateral contract and to take on their own respective counterparty risks. The CCP thus becomes the new counterparty for both parties in the contract, which are consequently only exposed to the counterparty risk of the CCP. This risk should, however, be small as the CCP is well capitalized and thus well equipped to face possible defaults of its members. In order to manage the counterparty risk of the members, the CCP relies on several mechanisms. The first one are participation constraints, which exclude counterparties with default probability above a certain acceptable threshold from dealing with the CCP. Upon initiation of a bilateral contract through the CCP, each party is required to post an initial margin to the CCP, usually in form of cash or highly liquid securities. This initial margin serves to cover most possible losses in case of default of a counterparty, whose positions are then inherited by the CCP. In particular, this concerns the period between the last time the defaulting party's position was valued and variation margins were paid and the close-out of the position. Variation margins represent a third mechanism of protection and concern the changes in the market value of counterparties' positions in the contract. The counterparty whose position has lost in value is obliged to post the variation margin to the CCP (which is necessarily done in cash as opposed to collateral agreements). The CCP typically passes this margin to the other counterparty. The variation margins are usually calculated and collected daily. Finally, as the fourth line of defense against losses due to defaults of its members, the CCP also disposes of a non-margin collateral such as default funds which contain collateral posted by all members of the CCP.

Acknowledging the role of collateralization and central clearing in mitigation of counterparty risk in OTC derivatives (we refer in particular to Cont et al. (2011) who found significant differences in the values of cleared and uncleared interest rate swaps), one still has to be aware of the remaining liquidity risk that might even get more pronounced as a consequence of need to finance the collateral postings and variation margin calls in adverse mark-to-markets conditions.

1.2.3   Clean Prices Versus Global Prices

The term "clean price" was introduced in the post-crisis literature in Crépey (2015) and Crépey et al. (2014) and refers to a price of an OTC derivative in a hypothetical situation where default and liquidity risk of the two counterparties are assumed to be negligible. In particular, in case of interest rate derivatives, this means that the counterparty and liquidity risk of the two parties are ignored, whereas the counterparty and liquidity risk of the interbank market, which directly influence the reference rates in these contracts and create the multiple curve phenomenon, are still taken into account. Hence, in this sense the interest rate derivative prices studied in this book are *clean, multiple curve* prices. This is also supported by the fact that most market quotations of derivative prices reflect collateralized transactions, thus leading to clean prices.

On the other side, a *global* price of a derivative is a price including also the adjustments due to counterparty and liquidity risk. This can be done in two ways: either by developing a pricing framework which takes these issues into account from the beginning, or by computing firstly the clean prices and then "adjusting" them for these risks by CVA (credit valuation adjustment), DVA (debt valuation adjustment) and FVA (funding valuation adjustment), as well as other adjustments referred to simply by XVA. The computation and the interplay of these adjustments in order to obtain the global price of a derivative are far from trivial.

Even though both of these approaches have their advantages and disadvantages for OTC pricing in general, in our view the second approach seems to be well suited for the case of interest rate derivatives. Firstly, the "splitting" of the pricing procedure into two parts corresponds in fact to common practice, where the clean prices are computed on the case by case basis (derivative by derivative) and then used as an underlying to produce the valuation adjustments, for which the whole portfolio between two counterparties, and not only one specific derivative, plays a role. Secondly, in contrast to credit derivatives, the wrong-way risk and the gap risk in interest rate derivatives are rather small, hence disregarding counterparty risk when computing clean interest rate derivative prices seems to be a reasonable assumption. Finally, since already the clean pricing of interest rate derivatives requires complex models due to the multiple curve issue, we feel that the two-step approach in obtaining the global derivative prices should be preferred in this case. The treatment in this book therefore concerns only the clean pricing of interest rate derivatives. For a detailed overview of the global pricing of financial derivatives we refer the interested reader to Brigo et al. (2013), Pallavicini and Brigo (2013), Crépey et al. (2014) and the references therein.

1.3 The New Paradigm: Multiple Curves at All Levels

Let us fix a finite time horizon for all market activities, denoted by $T^* > 0$. Having seen in the previous sections that the key role in the post-crisis fixed-income markets is played by the tenor of the underlying interest rate, let us now formalize these discussions and introduce the notation and the needed probabilistic framework.

A discrete *tenor structure* $\mathscr{T}^x$ with tenor x is a finite sequence of dates

$$\mathscr{T}^x := \{0 \leq T_0^x < T_1^x < \cdots < T_{M_x}^x \leq T^*\} \tag{1.9}$$

We denote $\delta_k^x := T_k^x - T_{k-1}^x$ the year fraction corresponding to the length of the interval $(T_{k-1}^x, T_k^x]$, for $k = 1, \ldots, M_x$. Typically, the distance between the dates in the tenor structure will be constant, i.e. $\delta_k^x = \delta^x$, for all k.

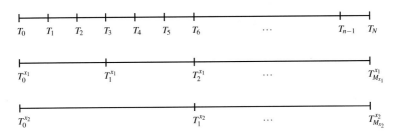

Fig. 1.5 Illustration of different tenor structures

Remark 1.1 In this book, for sake of clarity of the exposition, we put aside the practical issue of day count conventions. We acknowledge that in practice, however, there is a variety of day count conventions that have to be taken into account and refer to e.g. Ametrano and Bianchetti (2013) for more details on these conventions.

As already seen in the previous sections, in practice the tenor x ranges from one day ($\delta^x = \frac{1}{360}$) to twelve months ($\delta^x = 1$). In the multi-curve setup one has to consider different possible tenor structures simultaneously. We shall thus denote by $\mathscr{X} := \{x_1 < x_2 < \cdots < x_n\}$ a collection of tenors and by $\mathscr{T}^{x_i} = \{0 \leq T_0^{x_i} < \cdots < T_{M_{x_i}}^{x_i}\}$ the associated tenor structures for $i = 1, \ldots, n$, thereby assuming that $\mathscr{T}^{x_n} \subset \mathscr{T}^{x_{n-1}} \subset \cdots \subset \mathscr{T}^{x_1} \subseteq \mathscr{T}$, where $\mathscr{T} := \{0 \leq T_0 < T_1 < \cdots < T_M \leq T^*\}$ can be seen as a reference tenor structure containing all the others. Moreover, assume that $T_{M_{x_i}}^{x_i} = T_M$, for all i, meaning that all tenor structures have a common terminal date. Typically, we have $\mathscr{X} = \{1, 3, 6, 9, 12\}$ months. As an example, Fig. 1.5, taken from Grbac et al. (2014), illustrates the relation between different tenor dates in the 1-month, 3-month and 6-month tenor structures, assuming that the 1-month tenor structure is the reference tenor structure.

Having in mind the discussion at the beginning of this chapter and the impact of the underlying tenor on the values of the Libor rates, after the crisis, instead of having Libor rates of different tenors connected by no-arbitrage relations, one has to associate to each tenor $x \in \mathscr{X}$ a different curve. In other words, at time $t = 0$, for each x the following rates are observable, where here and below by "observable" we mean quantities that are either directly observable or can be computed from market data as will be explained further in Remark 1.2:

$$L(0; T_{k-1}^x, T_k^x), \qquad k = 1, \ldots, M_x \tag{1.10}$$

Hence, one can define the associated discount curve by imposing the following classical relation:

$$\frac{p^x(0, T_{k-1}^x)}{p^x(0, T_k^x)} := 1 + \delta_k^x L(0; T_{k-1}^x, T_k^x), \qquad k = 1, \ldots, M_x \tag{1.11}$$

and deriving from it the x-bond prices $p^x(0, T_k^x)$, cf. Ametrano and Bianchetti (2013) and Miglietta (2015). Note, as pointed out also by Miglietta (2015), that there is no unique inverse relationship between the initial Libor curve $L(0; T_{k-1}^x, T_k^x)$ and the initial x-bond term structure, only the quotients $\frac{p^x(0, T_{k-1}^x)}{p^x(0, T_k^x)}$ are uniquely determined. Clearly, the simultaneous presence of several, mutually "disconnected" Libor curves that cannot be associated to only one common discount curve $T \mapsto p(0, T)$, as it was the case in the pre-crisis setup, gives rise to the first obvious question of choice of the discount curve (or curves), as well as to other questions related to the mathematically sound and practically reasonable modeling of multiple curves.

Remark 1.2 (*Tradable quantities and market data*) Regarding the tradable quantities and bootstrapping of the yield curves based on market quotes, we briefly summarize the most important points from the paper by Ametrano and Bianchetti (2013), which provides a very detailed overview of the procedure of calibration and yield curve construction for the instruments traded in the European market. According to this paper, in the European market the most important tenors that are considered are the following ones: 1 day, 1, 3, 6 and 12 months and the most liquidly traded instruments are based on these tenors as underlyings. Constructions of the yield curves corresponding to the mentioned tenors is done by bootstrapping both from available market prices of traded instruments, as well as from prices of synthetic instruments, which are used for replacing the missing market quotes. These instruments include deposits (depos), FRAs, interest rate swaps, overnight indexed swaps and basis swaps based on various tenors. The most important yield curve is the one related to the 1-day tenor. This curve is constructed on the basis of the market OIS rates for overnight indexed swaps and is therefore usually called the OIS yield curve. More precisely, analogously to what is done in the pre-crisis setting, starting from the market OIS rates one can construct by bootstrapping the OIS bond prices (see Eq. (1.32) and Remark 1.6 in Sect. 1.4.4). This gives the OIS discount curve that in turn leads in the usual way to the OIS yield curve. The importance of the OIS discount curve (or equivalently the OIS yield curve) lies in the fact that it is the most commonly used discount curve for pricing of other interest rate derivatives. We recall, as mentioned already at the end of Sect. 1.1, that the underlying overnight rate of overnight indexed swaps in the European market is the Eonia rate.

The procedure of constructing the yield curves from market data is based in general on two types of algorithms: the *best-fit*, where a functional form for the yield curves, such as Nelson-Siegel or Svensson, is assumed and then the parameters are calibrated, and the *exact-fit*, where a number of pre-selected market instruments is repriced exactly by bootstrapping and then the interpolation is used to obtain the remaining maturities. Ametrano and Bianchetti (2013) also provide details on the baskets of instruments used in the construction of each of the yield curves.

In the subsequent sections we define various probability measures used for pricing of interest rate derivatives in the sequel. In order to do so, we introduce a filtered probability space $(\Omega, \mathscr{F}, (\mathscr{F}_t)_{0 \le t \le T^*}, Q)$, where the filtration $(\mathscr{F}_t)_{0 \le t \le T^*}$ is assumed to satisfy the usual conditions. All price processes introduced in the remainder of the chapter are defined on this probability space and adapted to the filtration $(\mathscr{F}_t)_{0 \le t \le T^*}$. We shall use the notation X, $(X_t)_{0 \le t \le T^*}$ or simply X_t to denote a stochastic process.

1.3.1 Choice of the Discount Curve

In the presence of multiple curves, the choice of the curve for discounting of the future cash flows, and a related choice of the standard martingale measure used for pricing (in other words, the question of absence of arbitrage), becomes non-trivial. One could possibly choose a different discounting curve depending on the tenor of the underlying interest rate and consider each x-tenor market as a separate market. However, note that this requires imposing in addition certain relations that ensure the absence of arbitrage between these markets that are interconnected by means of interest rate derivatives whose payments depend on more than one tenor simultaneously. The other possibility is to choose a common discounting curve that will apply to all future cash flows, regardless of their tenor. In fact, this is the choice that has been widely accepted and became practically standard, with the OIS discount curve (i.e. the discount curve stripped from the OIS rates) as the common discount curve, cf. also the comments in Remark 1.2. One of the main arguments justifying this choice, which is typically evoked, is the fact that in practice the majority of traded interest rate derivatives are nowadays being collateralized and the rate used for remuneration of the collateral is exactly the overnight rate, which is the rate the OIS are based on. Moreover, the overnight rate bears very little risk due to its short maturity and therefore can be considered relatively risk-free. For more detailed discussions on this issue we refer to Ametrano and Bianchetti (2013), Filipović and Trolle (2013) and Hull and White (2013).

A formal derivation of the OIS discount curve will be presented in Sect. 1.4.4. In the remainder of the book we shall assume that the discount curve is the OIS discount curve $T \mapsto p^{OIS}(t, T)$ for any t and, in order to simplify the notation, we shall just use $p(t, T)$. We shall call *OIS bonds* the, in general hypothetical, bonds with price $p(t, T) = p^{OIS}(t, T)$. These bonds are not necessarily traded since they correspond to OIS rates that are based on an averaging procedure.[5] In the literature they are however often assumed to be tradable assets as e.g. in Mercurio (2010a).

From the OIS bonds $p(t, T)$ we may formally derive corresponding instantaneous forward rates via the classical relationship

$$f(t, T) := -\frac{\partial}{\partial T} \log p(t, T) \tag{1.12}$$

and from here then obtain the spot rate $r_t = f(t, t)$ that we shall call the *OIS short rate*. In practice, this rate will be approximated by the overnight rate that corresponds to the shortest available tenor.

---

[5]We would like to thank Darrell Duffie for having clarified to us some issues pertaining to OIS bonds in relation to the standard traded bonds.

1.3.2 Standard Martingale Measure and Forward Measures Related to OIS Bonds

Given the OIS short rate r_t, we may define in the usual way the corresponding money market account as

$$B_t = \exp\left(\int_0^t r_s ds\right)$$

We consider as *standard martingale measure* a probability measure Q, equivalent to the physical measure P, under which all traded assets, discounted by B as numéraire, are (local) martingales.

By analogy to the classical bond price formula we now postulate for the OIS bonds the relationship

$$p(t, T) = E^Q\left\{\frac{B_t}{B_T}\bigg|\mathscr{F}_t\right\} = E^Q\left\{\exp\left[-\int_t^T r_s ds\right]\bigg|\mathscr{F}_t\right\} \tag{1.13}$$

which implies that the process $\left(\frac{p(t,T)}{B_t}\right)_{t\leq T}$ is, for each T, a Q-martingale. The meaningfulness of the above relationship between r_t and $p(t, T)$ stems from the fact that, whenever the OIS bonds are actually traded, their prices should be arbitrage-free. Formula (1.13) for $p(t, T)$, viewed as discount curve, can also be found in Kijima et al. (2009). Notice, furthermore, that formula (1.13) corresponds to formula (3), namely $P_c(t, T) = E^Q\left\{\exp\left[-\int_t^T r_c(s)ds\right] \mid \mathscr{F}_t\right\}$ in Filipović and Trolle (2013), which gives the price $P_c(t, T)$ of a collateralized zero coupon bond when the collateral rate r_c is the overnight rate (see also Piterbarg 2010 and Fujii et al. 2011). This is typically the case in practice as mentioned in Sect. 1.3.1. In this sense the OIS bonds correspond to collateralized zero coupon bonds that may actually be traded.

Since the process $\left(\frac{p(t,T)}{B_t}\right)_{t\leq T}$ is a Q-martingale, we can use it as density process for an equivalent measure change. In fact, we may now introduce the standard forward martingale measures defining, for a generic $T \in [0, T^*]$, the T-forward measure Q^T as given by

$$\frac{dQ^T}{dQ}\bigg|_{\mathscr{F}_t} = \frac{p(t, T)}{B_t}\frac{B_0}{p(0, T)} = \frac{p(t, T)}{B_t p(0, T)}, \qquad 0 \leq t \leq T \tag{1.14}$$

Note that the forward measure Q^T is associated to the OIS bond $p(\cdot, T)$ as numéraire, hence the density process is a ratio of the two numéraires. Moreover, the link between two forward measures associated to the dates $T, S \in [0, T^*]$ is given by

$$\frac{dQ^T}{dQ^S}\bigg|_{\mathscr{F}_t} = \frac{p(t, T)}{p(t, S)}\frac{p(0, S)}{p(0, T)}, \qquad 0 \leq t \leq T \wedge S \tag{1.15}$$

The forward martingale measures are particularly relevant in the context of interest
rate derivative pricing, but also for the direct modeling of the forward interest rates;
see the seminal paper by Geman et al. (1995), where the idea of changing a numéraire
(and thus changing a measure) has been first proposed in financial modeling as a tool
in asset pricing.

1.4 Interest Rate Derivatives

In this section an overview of the most standard interest rate derivatives is given
with precise definitions and connections between different derivatives. Moreover,
the quantities which may serve as building blocks for pricing models presented in
the sequel are identified.

A. Linear Derivatives
We begin by presenting the linear interest rate derivatives such as coupon bonds,
forward rate agreements and various types of interest rate swaps. A reader familiar
with the basics of the interest rate theory will know that before the crisis the prices
of these derivatives were given simply as linear combinations of zero coupon bond
prices. Now they are functions of both OIS bond prices and forward Libor rates.

1.4.1 Forward Rate Agreements

Definition 1.3 A *forward rate agreement* (FRA) is an OTC derivative that allows
the holder to lock in at any date $0 \leq t \leq T$ the interest rate between the inception
date T and the maturity $S > T$ at a fixed value R. At maturity S, a payment based on
R is made and the one based on the relevant floating rate *(generally the spot Libor
rate $L(T; T, S)$)* is received. The notional amount is denoted by N.

As discussed at the beginning of the chapter, before the crisis the FRA rate was
exactly the rate given by (1.4), where the last equality results from a no-arbitrage
argument in which one "locks-up" a rate between T and S by buying and selling
bonds of maturities T and S. Following a widely accepted practice in the post-crisis
literature, we now define the *post-crisis forward OIS rate*, based on the OIS bond
prices $p(t, T)$, as the discretely compounded forward rate given by

$$F(t; T, S) = \frac{1}{S - T} \left(\frac{p(t, T)}{p(t, S)} - 1 \right) \tag{1.16}$$

Notice that, although (1.16) coincides with the last expression in (1.4), the bonds there
are the pre-crisis zero coupon bonds, while here they are the OIS bonds. Sometimes,
the forward OIS rates $F(t; T, S)$ are also denoted by $L^D(t; T, S)$ to make explicit the
relation to the discount curve. In the pre-crisis framework, the forward Libor rate

$L(t; T, S)$ was assumed to be free of interbank risk and thus to coincide with the forward OIS rate, namely the following equality was supposed to hold

$$L(t; T, S) = L^D(t; T, S) = F(t; T, S)$$

Coming back to the FRA rates implied by FRA contracts on the spot Libor rate $L(T; T, S)$, recall that the spot Libor rate is no longer assumed to be free of various interbank risks and thus is no longer connected to the OIS bonds, i.e.

$$L(T; T, S) \neq \frac{1}{S - T} \left(\frac{1}{p(T, S)} - 1 \right)$$

The payoff of the FRA with notional amount N at maturity S is equal to

$$P^{FRA}(S; T, S, R, N) = N(S - T)(L(T; T, S) - R)$$

where $L(T; T, S)$ is the T-spot Libor rate for the time interval $[T, S]$. Thus, the value of the FRA at time $t \leq T$ is calculated as the conditional expectation with respect to the forward measure Q^S associated with the OIS bond with maturity S as numéraire and is given by

$$P^{FRA}(t; T, S, R, N) = N(S - T)p(t, S)E^{Q^S} \{L(T; T, S) - R | \mathscr{F}_t\} \quad (1.17)$$

Hence, the key quantity is the conditional expectation of the spot Libor rate that we denote by $L(t; T, S)$ and define by

$$L(t; T, S) := E^{Q^S} \{L(T; T, S) \,|\, \mathscr{F}_t\}, \quad 0 \leq t \leq T < S \quad (1.18)$$

As stated in Definition 1.2 we call this quantity the *forward Libor rate*, but we emphasize again the crucial difference with respect to the classical pre-crisis forward Libor rate, namely the connection to the bond prices which is now lost:

$$L(t; T, S) = E^{Q^S} \{L(T; T, S) \,|\, \mathscr{F}_t\} \neq \frac{1}{S - T} \left(\frac{p(t, T)}{p(t, S)} - 1 \right) \quad (1.19)$$

The value of the FRA at time t is then simply given by

$$P^{FRA}(t; T, S, R, N) = N(S - T)p(t, S)\,(L(t; T, S) - R) \quad (1.20)$$

and the forward rate R_t implied by this FRA at time $t \leq T$, i.e. the rate R such that $P^{FRA}(t; T, S, R, N) = 0$, is obviously equal to $L(t; T, S)$.

Remark 1.3 We mention here that the traded FRA contracts are in fact defined in a slightly different way. More precisely, the payoff of the market FRA (as opposed to the standard, textbook FRA defined above) is given by

$$P^{mFRA}(T; T, S, R, N) = N \frac{(S - T)(L(T; T, S) - R)}{1 + (S - T)L(T; T, S)}$$

$$= \frac{P^{FRA}(S; T, S, R, N)}{1 + (S - T)L(T; T, S)} \qquad (1.21)$$

and is paid at the beginning of the reference interval, i.e. at time T (in contrast to the payment at time S in the case of the standard FRA). Intuitively speaking, the payoff of the market FRA equals the payoff of the standard FRA paid at T instead of S, where the amount is discounted by a discount factor coming from the Libor curve $\frac{1}{1+(S-T)L(T;T,S)}$ (and not from the OIS curve). Obviously, in the pre-crisis setup the market and the standard FRA definitions were equivalent, as it can be easily checked by a simple calculation.

In the sequel, when writing only FRA, we shall always mean the standard FRA. The difference in definitions should be kept in mind when calibrating a model to market data, although it has been pointed out by e.g. Mercurio (2010b) that the actual difference in value of the contract is small enough to be neglected.

Remark 1.4 The forward Libor rate together with the OIS bond prices are the building blocks for the prices of other linear interest rate derivatives such as various types of swaps. Many models in the recent literature consider either directly the forward Libor rate, or one of the related spreads, as a modeling object. These are the models in the spirit of the classical Libor market models, which are treated in Chap. 4. Another approach is to focus on the pre-crisis connection of the spot Libor rates and bond prices and introduce the following relation:

$$L(T; T, S) = \frac{1}{S - T}\left(\frac{1}{\bar{p}(T, S)} - 1\right) \qquad (1.22)$$

and set according to Definition 1.2

$$L(t; T, S) = E^{Q^S}\left\{\frac{1}{S - T}\left(\frac{1}{\bar{p}(T, S)} - 1\right)\Big|\mathscr{F}_t\right\} \qquad (1.23)$$

where $\bar{p}(T, S)$ can be interpreted as price of a fictitious risky bond that is supposed to be affected by the same risk factors as the Libor rate. Here, we kept the classical formal relationship between the Libor rates and the bond prices, but replaced the prices $p(T, S)$ in the classical relationship by the prices $\bar{p}(T, S)$ of the fictitious bonds. Note that these fictitious bonds are not traded assets, but can be considered as being issued by an average Libor bank, see Ametrano and Bianchetti (2013) and Morini (2009). This is why these bonds are referred to as the Libor bonds by some authors. In Gallitschke et al. (2014) they are called interbank bonds, since interbank cash transactions can be represented as interbank bonds. The models for these bonds are then specified by specifying the dynamics of the process $(\bar{p}(t, S))_{0 \leq t \leq S}$ either directly (HJM approach, see Chap. 3), or via a suitable short-rate process (for the short-rate approach see Chap. 2).

1.4.2 Fixed and Floating Rate Bonds

Definition 1.4 A *fixed rate bond (fixed rate note)* is a financial instrument offering to its holder a stream of future payments called *coupons*. Denoting by $0 \leq T_0 < T_1 < \cdots < T_n$ a discrete tenor structure with $\delta_k = T_k - T_{k-1}$ and by N the notional amount, the fixed rate bond pays out the amount $N\delta_k c_k$ at date T_k, for $c_k \in (0,1)$ and $k = 1, \ldots n$. The notional amount N is paid in addition to the coupon payment at maturity T_n. In a *floating rate bond (floating rate note)* the coupon payments are based on a floating rate (generally the spot Libor rate for a given period), i.e. the floating rate bond pays out the amount $N\delta_k L(T_{k-1}; T_{k-1}, T_k)$ at date T_k, where $L(T_{k-1}; T_{k-1}, T_k)$ is the spot Libor rate fixed at T_{k-1} for the period $[T_{k-1}, T_k]$ with $k = 1, \ldots n$. In the sequel we shall use the shorthand notation $Q^k = Q^{T_k}$ for the forward measures.

The price at time $t \leq T_0$ of the fixed rate bond is given by

$$p^c(t, T_n) = \sum_{k=1}^{n} N\delta_k p(t, T_k) c_k + Np(t, T_n) \tag{1.24}$$

Similarly, the price of the floating rate bond at time $t \leq T_0$ can be expressed as

$$p^{float}(t, T_n) = \sum_{k=1}^{n} N\delta_k p(t, T_k) E^{Q^k}\{L(T_{k-1}; T_{k-1}, T_k)|\mathscr{F}_t\} + Np(t, T_n)$$

$$= \sum_{k=1}^{n} N\delta_k p(t, T_k) L(t; T_{k-1}, T_k) + Np(t, T_n) \tag{1.25}$$

We recall that, before the crisis, the price of the floating rate bond was simply

$$p^{float}(t, T_n) = \sum_{k=1}^{n} N\delta_k p(t, T_k) L(t; T_{k-1}, T_k) + Np(t, T_n)$$

$$= \sum_{k=1}^{n} N\delta_k p(t, T_k) \frac{1}{\delta_k}\left(\frac{p(t, T_{k-1})}{p(t, T_k)} - 1\right) + Np(t, T_n)$$

$$= Np(t, T_0) \tag{1.26}$$

due to the pre-crisis connection between the forward Libor rates and the bond prices as specified in (1.6). The third equality follows by cancellations in the telescopic sum. This means that the spot starting floating rate bond was worth *par*, i.e. $p^{float}(T_0, T_n) = N$.

1.4.3 Interest Rate Swaps

In full generality, a swap is a financial contract between two parties to exchange one
stream of future payments for another one.

Definition 1.5 An *interest rate swap* is a financial contract in which a stream of
future interest rate payments linked to a pre-specified fixed rate denoted by R is
exchanged for another one linked to a floating interest rate, based on a specified
notional amount N. The floating rate is generally taken to be the Libor rate, with
various possible conventions concerning the fixing and the payment dates. The swap
is initiated at time $T_0 \geq 0$ and $T_1 < \cdots < T_n$, where $T_1 > T_0$, denote a collection of
the payment dates, with $\delta_k := T_k - T_{k-1}$, for all $k = 1, \ldots, n$.

Note that in this book we shall always use the convention where the floating rates
are fixed in advance and the payments are made in arrears. Moreover, note also that
there exist other possible choices for the floating rate besides the Libor rates. One
such example is treated below, namely the overnight indexed swap (OIS), in which
the floating rate is obtained by compounding the overnight rates. Further examples
include constant maturity swaps, in which the floating rates are market swap rates of
Libor-indexed swaps.

We recall that if the fixed rate is paid and the floating rate is received, the swap is
called a *payer swap*, as opposed to a *receiver swap*, where the fixed rate is received
and the floating rate is paid. If not specified otherwise, we shall always consider a
payer swap. The time-t value of the swap, where $t \leq T_0$, is given as a difference of
the time-t values of the floating leg and the fixed leg and is equal to

$$P^{Sw}(t; T_0, T_n, R, N) = N \sum_{k=1}^{n} \delta_k p(t, T_k) E^{Q^k} \{L(T_{k-1}; T_{k-1}, T_k) - R | \mathscr{F}_t\}$$

$$= N \sum_{k=1}^{n} P^{FRA}(t; T_{k-1}, T_k, R, 1)$$

$$= N \sum_{k=1}^{n} \delta_k p(t, T_k) (L(t; T_{k-1}, T_k) - R) \qquad (1.27)$$

where $L(t; T_{k-1}, T_k)$ is given by (1.18), for every $k = 1, \ldots, n$. The swap rate
$R(t; T_0, T_n)$ is the rate that makes the time-t value $P^{Sw}(t; T_0, T_n, R, N)$ of the swap
equal to zero and it is easily seen that it is given by

$$R(t; T_0, T_n) = \frac{\sum_{k=1}^{n} \delta_k p(t, T_k) L(t; T_{k-1}, T_k)}{\sum_{k=1}^{n} \delta_k p(t, T_k)} \qquad (1.28)$$

$$= \sum_{k=1}^{n} w_k L(t; T_{k-1}, T_k)$$

i.e. the swap rate is a convex combination of the forward Libor rates, with weights
$w_k := \frac{\delta_k p(t, T_k)}{\sum_{i=1}^{n} \delta_i p(t, T_i)}$, $k = 1, \ldots, n$, which are functions of the OIS bond prices.

Remark 1.5 In practice the floating rate payments and the fixed rate payments of the swap defined above typically do not occur with the same frequency, as we have assumed to simplify the notation. For example, in the European markets, the fixed leg payments typically occur on a one-year tenor structure, whereas the floating rate payments adopt the tenor of the underlying Libor rate (from one month to three months and up to one year). In that case one has to work with two different tenor structures and modify the above formulas accordingly. In particular, if we denote by $\mathscr{T}^x$ the tenor structure for the floating rate payments and by $\mathscr{T}^y$ the tenor structure for the fixed rate payments, the time-t value of this interest rate swap is given, with some abuse of notation concerning the symbol P^{Sw} for the swap value, by

$$P^{Sw}(t; \mathscr{T}^x, \mathscr{T}^y, R, N) = N \left(\sum_{i=1}^{n_x} \delta_i^x p(t, T_i^x) L(t; T_{i-1}^x, T_i^x) - R \sum_{j=1}^{n_y} \delta_j^y p(t, T_j^y) \right)$$

which follows exactly by the same reasoning as (1.27). The corresponding swap rate is given by

$$R(t; \mathscr{T}^x, \mathscr{T}^y) = \frac{\sum_{i=1}^{n_x} \delta_i^x p(t, T_i^x) L(t; T_{i-1}^x, T_i^x)}{\sum_{j=1}^{n_y} \delta_j^y p(t, T_j^y)}$$

1.4.4 Overnight Indexed Swaps (OIS)

In an overnight indexed swap (OIS) the counterparties exchange a stream of fixed rate payments for a stream of floating rate payments linked to a compounded overnight rate. Let us assume the same tenor structure as in the previous subsection is given and denote again the fixed rate by R. The time-t value $P^{OIS}(t; T_0, T_n, R, N)_{fix}$ of the fixed leg payments is given by

$$P^{OIS}(t; T_0, T_n, R, N)_{fix} = NR \sum_{k=1}^{n} \delta_k p(t, T_k),$$

whereas to compute the value of the floating leg $P^{OIS}(t; T_0, T_n, R, N)_{float}$ we proceed as follows: the floating rate for each interval $(T_{k-1}, T_k]$ is given by simply compounding the overnight rates between these two dates, i.e.

$$R^{ON}(T_{k-1}, T_k) = \frac{1}{\delta_k} \left(\prod_{j=1}^{n_k} \left[1 + \delta_{t_{j-1}^k, t_j^k} R^{ON}(t_{j-1}^k, t_j^k) \right] - 1 \right)$$

where $T_{k-1} = t_0^k < t_1^k < \cdots < t_{n_k}^k = T_k$ is the subdivision into dates of the fixings of the overnight rate (i.e. working days) and $\delta_{t_{j-1}^k, t_j^k} := t_j^k - t_{j-1}^k$, with $R^{ON}(t_{j-1}^k, t_j^k)$

denoting thus the overnight rate for the period $(t_{j-1}^k, t_j^k]$. The payment based on the discretely compounded rate $R^{ON}(T_{k-1}, T_k)$ is made at T_k. In order to find the value at time t of this payment, we proceed with a calculation inspired by the one in Ametrano and Bianchetti (2013). The overnight rate $R^{ON}(t_{j-1}^k, t_j^k)$ is assumed to be linked to the OIS bond prices via the classical pre-crisis forward rate formula

$$R^{ON}(t_{j-1}^k, t_j^k) = \frac{1}{\delta_{t_{j-1}^k, t_j^k}} \left(\frac{p(t_{j-1}^k, t_{j-1}^k)}{p(t_{j-1}^k, t_j^k)} - 1 \right) \tag{1.29}$$

We refer to Filipović and Trolle (2013, Sect. 2.5) for a derivation of the above formula based on continuous compounding of the instantaneous rate approximating the overnight rate. Hence, the value of the floating leg is given by

$$P^{OIS}(t; T_0, T_n, R, N)_{float} = N \sum_{k=1}^{n} \delta_k p(t, T_k) R^{ON}(t; T_{k-1}, T_k)$$

where

$$
\begin{aligned}
R^{ON}(t; T_{k-1}, T_k) &= E^{Q_{T_k}} \left\{ R^{ON}(T_{k-1}, T_k) | \mathscr{F}_t \right\} \\
&= \frac{1}{\delta_k} E^{Q_{T_k}} \left\{ \left(\prod_{j=1}^{n_k} \left[1 + \delta_{t_{j-1}^k, t_j^k} R^{ON}(t_{j-1}^k, t_j^k) \right] - 1 \right) \Big| \mathscr{F}_t \right\} \\
&= \frac{1}{\delta_k} \left(E^{Q_{T_k}} \left\{ \prod_{j=1}^{n_k} \frac{p(t_{j-1}^k, t_{j-1}^k)}{p(t_{j-1}^k, t_j^k)} \Big| \mathscr{F}_t \right\} - 1 \right) \\
&= \frac{1}{\delta_k} \left(\frac{p(t, T_{k-1})}{p(t, T_k)} - 1 \right)
\end{aligned}
\tag{1.30}
$$

The third equality follows from (1.29) and the fourth one is based on a sequence of subsequent measure changes from Q_{T_k} to $Q_{t_j^k}$, for $j = n_k - 1, \ldots, 0$ (recall that $T_k = t_{n_k}^k$). To be more precise, in the first step, making use of the density between the forward measures $Q_{t_{n_k}^k}$ and $Q_{t_{n_k-1}^k}$ given in (1.15) and applying the abstract Bayes rule, we have

$$
\begin{aligned}
E^{Q_{T_k}} \left\{ \prod_{j=1}^{n_k} \frac{p(t_{j-1}^k, t_{j-1}^k)}{p(t_{j-1}^k, t_j^k)} \Big| \mathscr{F}_t \right\} &= \frac{E^{Q_{t_{n_k-1}^k}} \left\{ \prod_{j=1}^{n_k-1} \frac{p(t_{j-1}^k, t_{j-1}^k)}{p(t_{j-1}^k, t_j^k)} \Big| \mathscr{F}_t \right\}}{E^{Q_{t_{n_k-1}^k}} \left\{ \frac{p(t_{n_k-1}^k, t_{n_k}^k)}{p(t_{n_k-1}^k, t_{n_k-1}^k)} \Big| \mathscr{F}_t \right\}} \\
&= \frac{p(t, t_{n_k-1}^k)}{p(t, t_{n_k}^k)} E^{Q_{t_{n_k-1}^k}} \left\{ \prod_{j=1}^{n_k-1} \frac{p(t_{j-1}^k, t_{j-1}^k)}{p(t_{j-1}^k, t_j^k)} \Big| \mathscr{F}_t \right\}
\end{aligned}
$$

Repeating the same procedure, we obtain the following telescopic product

$$E^{Q_{T_k}}\left\{\prod_{j=1}^{n_k}\frac{p(t_{j-1}^k, t_{j-1}^k)}{p(t_{j-1}^k, t_j^k)}\bigg|\mathscr{F}_t\right\} = \prod_{j=1}^{n_k}\frac{p(t, t_{j-1}^k)}{p(t, t_j^k)} = \frac{p(t, T_{k-1})}{p(t, T_k)}$$

which concludes the derivation of (1.30). Consequently,

$$P^{OIS}(t; T_0, T_n, R, N)_{float} = N\sum_{k=1}^{n}\delta_k p(t, T_k)\frac{1}{\delta_k}\left(\frac{p(t, T_{k-1})}{p(t, T_k)} - 1\right)$$
$$= N\left(p(t, T_0) - p(t, T_n)\right)$$

where the second equality follows by cancellations in the telescopic sum.

Therefore, the time-t value, for $t \leq T_0$, of the payer OIS (i.e. the OIS in which the floating rate is received and the fixed rate is paid) is given by

$$P^{OIS}(t; T_0, T_n, R, N) = N\left(p(t, T_0) - p(t, T_n) - R\sum_{k=1}^{n}\delta_k p(t, T_k)\right) \quad (1.31)$$

The OIS rate $R^{OIS}(t; T_0, T_n)$, for $t \leq T_0$, is the rate R such that the value of the OIS at time t is equal to zero, i.e. $P^{OIS}(t; T_0, T_n, R, N) = 0$. It is given by

$$R^{OIS}(t; T_0, T_n) = \frac{p(t, T_0) - p(t, T_n)}{\sum_{k=1}^{n}\delta_k p(t, T_k)} \quad (1.32)$$

and coincides with the classical pre-crisis swap rate, compare also Filipović and Trolle (2013, Sect. 2.5, Eq. 11).

Remark 1.6 Note that the OIS discount curve is obtained by stripping the OIS bond prices based on the expression (1.32) for the OIS rate, see Remark 1.2. More precisely, assuming a collection of market quotes for OIS rates of overnight indexed swaps with various lengths is given, the relationship (1.32) allows to obtain the OIS bond prices $p(t, T_k)$ by solving a corresponding system of equations.

The *additive spot Libor-OIS spread* at time T, for the interval $[T, T + \Delta]$, where $T \geq 0$ and $\Delta > 0$, is thus given by

$$S(T; T, T + \Delta) := L(T; T, T + \Delta) - R^{OIS}(T; T, T + \Delta) \quad (1.33)$$
$$= L(T; T, T + \Delta) - \frac{1}{\Delta}\left(\frac{1}{p(T, T + \Delta)} - 1\right)$$

where we have used (1.32) with a single payment date. Note that since a swap with a single payment date is in fact an FRA, we have $R^{OIS}(T; T, T+\Delta) = F(T; T, T+\Delta)$, where $F(T; T, T + \Delta)$ is a discretely compounded forward rate from Eq. (1.16). Even though it is not directly observable in the market, for modeling purposes, a *multiplicative spot Libor-OIS spread* sometimes turns out to be more convenient:

$$\Sigma(T; T, T + \Delta) := \frac{1 + \Delta L(T; T, T + \Delta)}{1 + \Delta R^{OIS}(T; T, T + \Delta)} \tag{1.34}$$

see Henrard (2014) and Cuchiero et al. (2015). Note here that the quantity in the numerator above is exactly the inverse of the Libor discount factor mentioned in connection to market FRAs in Remark 1.3. Therefore, multiplying the payoff of the market FRA $P^{mFRA}(T; T, S, R, N)$ by the multiplicative Libor-OIS spread $\Sigma(T; T, T + \Delta)$ one can express it as a payoff of the standard FRA discounted with an OIS discount factor.

Similarly, the *additive forward Libor-OIS spread* at time $t \leq T$, for the interval $[T, T + \Delta]$, is given by

$$S(t; T, T + \Delta) := L(t; T, T + \Delta) - R^{OIS}(t; T, T + \Delta) \tag{1.35}$$

$$= L(t; T, T + \Delta) - \frac{1}{\Delta} \left(\frac{p(t, T)}{p(t, T + \Delta)} - 1 \right)$$

and the *multiplicative forward Libor-OIS spread* is

$$\Sigma(t; T, T + \Delta) := \frac{1 + \Delta L(t; T, T + \Delta)}{1 + \Delta R^{OIS}(t; T, T + \Delta)} \tag{1.36}$$

The *Libor-OIS swap spread* at time $t \in [0, T_0]$ is by definition the difference between the swap rate (1.28) of the Libor-indexed interest rate swap and the OIS rate (1.32) and is given by

$$R(t; T_0, T_n) - R^{OIS}(t; T_0, T_n) = \frac{\sum_{k=1}^{n} \delta_k p(t, T_k) L(t; T_{k-1}, T_k) - p(t, T_0) + p(t, T_n)}{\sum_{k=1}^{n} \delta_k p(t, T_k)} \tag{1.37}$$

Remark 1.7 Similarly to Remark 1.5, in practice the floating rate payments and the fixed rate payments of the OIS take place on different tenor structures. Hence, if we denote by $\mathscr{T}^x$ the tenor structure for the floating rate payments and by $\mathscr{T}^y$ the tenor structure for the fixed rate payments, the time-t value of this OIS is given by

$$P^{OIS}(t; \mathscr{T}^x, \mathscr{T}^y, R, N) = N \left(p(t, T_0) - p(t, T_{n_x}^x) - R \sum_{k=1}^{n_y} \delta_k^y p(t, T_k^y) \right)$$

which follows exactly by the same reasoning as (1.31). The corresponding swap rate is given by

$$R^{OIS}(t; \mathscr{T}^x, \mathscr{T}^y) = \frac{p(t, T_0) - p(t, T_{n_x}^x)}{\sum_{k=1}^{n_y} \delta_k^y p(t, T_k^y)}$$

1.4.5 Basis Swaps

A basis swap is an interest rate swap, where two floating payments linked to the Libor rates of different tenors are exchanged. For example, a buyer of such a swap receives semiannually a 6m-Libor and pays quarterly a 3m-Libor, both set in advance and paid in arrears. Note that there also exist other conventions regarding the payments on the two legs of a basis swap. Below we give a definition of a generic basis swap.

Definition 1.6 A *basis swap* is a financial contract to exchange two streams of payments based on the floating rates (typically Libor rates) linked to two different tenor structures denoted by $\mathcal{T}^1 = \{T_0^1 < \cdots < T_{n_1}^1\}$ and $\mathcal{T}^2 = \{T_0^2 < \cdots < T_{n_2}^2\}$, where $T_0^1 = T_0^2 \geq 0$, $T_{n_1}^1 = T_{n_2}^2$, and $\mathcal{T}^1 \subset \mathcal{T}^2$. The notional amount is denoted by N, $T_0^1 = T_0^2$ is called the initiation date, $T_{n_1}^1 = T_{n_2}^2$ the maturity date of the basis swap and the first payments are due at T_1^1 and T_1^2, respectively.

The time-t value of the basis swap is, for $t \leq T_0^1 = T_0^2$, given by

$$P^{BSw}(t; \mathcal{T}^1, \mathcal{T}^2, N) = N\left(\sum_{i=1}^{n_1} \delta_i^1 p(t, T_i^1) E^{Q^{T_i^1}}\left\{L(T_{i-1}^1; T_{i-1}^1, T_i^1)|\mathcal{F}_t\right\} \right.$$
$$\left. - \sum_{j=1}^{n_2} \delta_j^2 p(t, T_j^2) E^{Q^{T_j^2}}\left\{L(T_{j-1}^2; T_{j-1}^2, T_j^2)|\mathcal{F}_t\right\}\right)$$

(1.38)

Thus, we have

$$P^{BSw}(t; \mathcal{T}^1, \mathcal{T}^2, N) = N\left(\sum_{i=1}^{n_1} \delta_i^1 p(t, T_i^1) L(t; T_{i-1}^1, T_i^1) \right.$$
$$\left. - \sum_{j=1}^{n_2} \delta_j^2 p(t, T_j^2) L(t; T_{j-1}^2, T_j^2)\right)$$

(1.39)

where $L(t; T_{k-1}^x, T_k^x)$ is given by (1.18), for each tenor structure $\mathcal{T}^x$, $x = 1, 2$ and $k = 1, \ldots, n_x$.

Note that in the classical one-curve setup the time-t value of such a swap is zero, whereas since the crisis markets quote positive basis swap spreads that have to be added to the payments made on the smaller tenor leg. More precisely, recalling that the smaller tenor leg corresponds to $\mathcal{T}^2$, the floating interest rate $L(T_{j-1}^2; T_{j-1}^2, T_j^2)$ at T_j^2 is replaced by $L(T_{j-1}^2; T_{j-1}^2, T_j^2) + S$, for every $j = 1, \ldots, n_2$, where S is the basis swap spread. The value of the basis swap with the added spread S is denoted by $P^{BSw}(t; \mathcal{T}^1, \mathcal{T}^2, S, N)$ and is given by an expression analogous to (1.39), which follows from (1.38), where $L(T_{j-1}^2; T_{j-1}^2, T_j^2)$ is replaced by $L(T_{j-1}^2; T_{j-1}^2, T_j^2) + S$, for every j. The fair basis swap spread $S^{BSw}(t; \mathcal{T}^1, \mathcal{T}^2)$ at the time t when the swap is contracted is the spread S which makes the t-value of the swap equal to zero, i.e. it results from solving $P^{BSw}(t; \mathcal{T}^1, \mathcal{T}^2, S, N) = 0$ and is given by

$$S^{BSw}(t; \mathcal{T}^1, \mathcal{T}^2) = \frac{\sum_{i=1}^{n_1} \delta_i^1 p(t, T_i^1) L(t; T_{i-1}^1, T_i^1) - \sum_{j=1}^{n_2} \delta_j^2 p(t, T_j^2) L(t; T_{j-1}^2, T_j^2)}{\sum_{j=1}^{n_2} \delta_j^2 p(t, T_j^2)}$$

$$(1.40)$$

We may check that the value of the basis swap in a pre-crisis one-curve setup is indeed zero. To this purpose recall first that in this setup the pre-crisis forward Libor rates, which in (1.6) were defined using the risk-free zero coupon bonds as $\left(L(t; T, T + \Delta) = \frac{1}{\Delta}\left(\frac{p(t,T)}{p(t,T+\Delta)} - 1\right)\right)_{0 \le t \le T}$, are martingales under the corresponding forward measures. We thus have

$$P^{BSw}(t; \mathcal{T}^1, \mathcal{T}^2, N) = N\left(\sum_{i=1}^{n_1} \delta_i^1 p(t, T_i^1) E^{Q^{T_i^1}}\left\{ L(T_{i-1}^1; T_{i-1}^1, T_i^1)|\mathcal{F}_t\right\}\right.$$

$$\left. - \sum_{j=1}^{n_2} \delta_j^2 p(t, T_j^2) E^{Q^{T_j^2}}\left\{ L(T_{j-1}^2; T_{j-1}^2, T_j^2)|\mathcal{F}_t\right\} \right)$$

$$= N\left(\sum_{i=1}^{n_1} \delta_i^1 p(t, T_i^1) L(t; T_{i-1}^1, T_i^1)\right.$$

$$\left. - \sum_{j=1}^{n_2} \delta_j^2 p(t, T_j^2) L(t; T_{j-1}^2, T_j^2)\right)$$

$$= N\left((p(t, T_0^1) - p(t, T_{n_1}^1)) - (p(t, T_0^2) - p(t, T_{n_2}^2))\right) = 0$$

by the assumptions $T_0^1 = T_0^2$ and $T_{n_1}^1 = T_{n_2}^2$.

In the multiple curve setup we cannot use the same calculation, since now the Libor rates are not connected to the bond prices as above. Hence, one ends up with formula (1.39), which in general yields a non-zero value of the basis swap and produces a non-zero basis swap spread (1.40). The market spreads are typically positive, hence multiple curve models are usually constructed in such a way that ensures this property of the model spreads (1.40).

Remark 1.8 Note that the price of a basis swap $P^{BSw}(t; \mathcal{T}^1, \mathcal{T}^2)$ with tenor structures $\mathcal{T}^1$ and $\mathcal{T}^2$ can be expressed as a difference of prices of two interest rate swaps which share the same tenor structure $\mathcal{T}^3$ for the fixed rate payments and the same fixed rate R, and the floating rate payments are made respectively on the two tenor structures $\mathcal{T}^1$ and $\mathcal{T}^2$ of the basis swap. More precisely, we have

$$P^{BSw}(t; \mathcal{T}^1, \mathcal{T}^2, N) = P^{Sw}(t; \mathcal{T}^1, \mathcal{T}^3, R, N) - P^{Sw}(t; \mathcal{T}^2, \mathcal{T}^3, R, N) \quad (1.41)$$

where $P^{Sw}(t; \mathcal{T}^1, \mathcal{T}^3, R, N)$ and $P^{Sw}(t; \mathcal{T}^1, \mathcal{T}^3, R, N)$ are defined as in Remark 1.5. Clearly, one could take the tenor structure of the fixed leg to coincide with the tenor structure of one of the floating legs, but, as mentioned already, in practice the fixed leg is often paid on a one-year tenor structure (at least in the European markets) and is thus common for both interest rate swaps, but the floating payments are made on two different tenor structures in general. Finally, we want to point out that, based on the representation (1.41), we could follow a slightly different convention concerning the basis swap spread (see Ametrano and Bianchetti 2013), namely by defining

$$S^{BSw}(t; \mathcal{T}^1, \mathcal{T}^2) = R(t; \mathcal{T}^1, \mathcal{T}^3) - R(t; \mathcal{T}^2, \mathcal{T}^3)$$

where $R(t; \mathcal{T}^i, \mathcal{T}^3)$, $i = 1, 2$, is the swap rate as defined in Remark 1.5. Notice that the difference with the definition in (1.40) consists in the denominator which, in the new formulation, stems from the fixed leg with tenor $\mathcal{T}^3$.

B. Optional Derivatives

The most common nonlinear interest rate derivatives are caps, floors and swaptions. All these derivatives are of optional nature, therefore we refer to them as *optional derivatives*. Caps (floors) consist of a series of call (put) options on a floating interest rate and swaptions are options which allow to enter into an underlying interest rate swap of the various types described in the previous part A.

1.4.6 Caps and Floors

Recall that an interest rate cap, respectively floor, is a optional financial derivative defined on a pre-specified discrete tenor structure. The buyer of a cap, respectively floor, has a right to payments at the end of each sub-period in the tenor structure in which the interest rate exceeds, respectively falls below, a mutually agreed strike level K. These payments, made by the seller, are thus given as a positive part of the difference between the interest rate and the strike K for the cap, respectively a positive part of the difference between the strike K and the interest rate for the floor. Every cap, respectively floor, can be decomposed into a series of options applying to each sub-period, which are called caplets, respectively floorlets. Below we thus focus on a single caplet with generic inception date and maturity, whereas the reasoning for a floorlet is completely symmetric.

Definition 1.7 A *caplet* with strike K, inception date $T \geq 0$ and maturity date $T + \Delta$, with $\Delta > 0$, on a nominal N is a financial contract whose holder has the right to a payoff at maturity given by $N\Delta(L(T; T, T + \Delta) - K)^+$, where $(L(T; T, T + \Delta)$ is the spot Libor rate fixed at time T for the time interval $[T, T + \Delta]$.

Note that the caplet can be seen as a call option with maturity $T + \Delta$ and strike K on the Libor rate, where the Libor rate is known already at time T, i.e. the caplet is said to be settled in arrears (again, similarly to Sect. 1.4.3, we adopt one of several possible settlement conventions). In the sequel we shall assume without loss of generality that the nominal is equal to one, i.e. $N = 1$.

The time-t price, for $t \leq T$, of the caplet is given by

$$P^{Cpl}(t; T + \Delta, K) = \Delta\, p(t, T + \Delta) E^{Q^{T+\Delta}} \left\{ (L(T; T, T + \Delta) - K)^+ \mid \mathscr{F}_t \right\}$$

In case of (1.22), we further have

$$P^{Cpl}(t; T + \Delta, K) = p(t, T + \Delta) E^{Q^{T+\Delta}} \left\{ \left(\frac{1}{\bar{p}(T, T + \Delta)} - \bar{K} \right)^+ \bigg| \mathscr{F}_t \right\}$$

$$(1.42)$$

where $\bar{K} = 1 + \Delta K$.

It is worthwhile mentioning that, when using the representation (1.22), the classical transformation of a caplet into a put option on a bond does not work in the multiple curve setup. More precisely, the fact that the payoff $\left((1 + \Delta L(T; T, T + \Delta)) - \bar{K} \right)^+$ at time $T + \Delta$ is equivalent to the payoff $p(T, T + \Delta) \left((1 + \Delta L(T; T, T + \Delta)) - \bar{K} \right)^+$ at time T is still valid, since the OIS discounting is used. However, this will not yield the desired cancellation of discount factors. Since the Libor rate depends on the fictitious bonds $\bar{p}(T, T + \Delta)$ and the OIS bonds $p(T, T + \Delta)$ are used for discounting, we have

$$p(T, T + \Delta) \left((1 + \Delta L(T; T, T + \Delta)) - \bar{K} \right)^+ = p(T, T + \Delta) \left(\frac{1}{\bar{p}(T, T + \Delta)} - \bar{K} \right)^+$$

which cannot be simplified further as in the one-curve case.

1.4.7 Swaptions

Definition 1.8 Consider a generic fixed-for-floating (payer) interest rate swap with inception date T_0, maturity date T_n and nominal N. A *swaption* is an option to enter the underlying swap at a pre-specified swap rate R, called the swaption strike rate, and a pre-specified date $T \leq T_0$ called the maturity of the swaption.

Let us consider the Libor-indexed interest rate swap from Sect. 1.4.3 and assume that the notional amount is $N = 1$. Moreover, for simplicity we choose the maturity of the swaption to coincide with the starting date of the swap, i.e. $T = T_0$. Therefore, the payoff of the swaption at maturity is given by $\left(P^{Sw}(T_0; T_0, T_n, R) \right)^+$ and we shall use the shorthand notation $P^{Sw}(T_0; T_n, R) = P^{Sw}(T_0; T_0, T_n, R)$. The value $P^{Swn}(t; T_0, T_n, R)$ of the swaption at time $t \leq T_0$ is

$$P^{Swn}(t; T_0, T_n, R) = p(t, T_0)E^{Q^{T_0}}\left\{\left(P^{Sw}(T_0; T_n, R)\right)^+ | \mathscr{F}_t\right\}$$

$$= p(t, T_0)E^{Q^{T_0}}\left\{\left(\sum_{k=1}^{n} \delta_k p(T_0, T_k)L(T_0; T_{k-1}, T_k)\right.\right.$$

$$\left.\left. -R\sum_{k=1}^{n} \delta_k p(T_0, T_k)\right)^+ \Bigg| \mathscr{F}_t\right\} \tag{1.43}$$

$$= p(t, T_0)E^{Q^{T_0}}\left\{\sum_{k=1}^{n} \delta_k p(T_0, T_k)\left(R(T_0; T_0, T_n) - R\right)^+ \Bigg| \mathscr{F}_t\right\}$$

$$= p(t, T_0)\sum_{k=1}^{n} \delta_k E^{Q^{T_0}}\left\{p(T_0, T_k)\left(R(T_0; T_0, T_n) - R\right)^+ | \mathscr{F}_t\right\}$$

The second equality follows from (1.27) and the third one from (1.28), where $R(T_0; T_0, T_n)$ is the swap rate of the underlying swap at time T_0. Note that the last equality allows to perceive a swaption as a sequence of payments $\delta_k \left(R(T_0; T_0, T_n) - R\right)^+$, $k = 1, \ldots, n$, fixed at time T_0, that are received at payment dates $T_1, \ldots, T_n$. These payments are equivalent to the payments $p(T_0, T_k)$ $\delta_k \left(R(T_0; T_0, T_n) - R\right)^+$, $k = 1, \ldots, n$, received at T_0, cf. Musiela and Rutkowski (2005, Sect. 13.1.2, p. 482).

To price a swaption, it is convenient to introduce the following process

$$A_t := \sum_{k=1}^{n} \delta_k p(t, T_k), \qquad t \leq T_1$$

Because A_t is a linear combination of OIS prices, the process $\left(\frac{A_t}{p(t, T_0)}\right)_{t \leq T_0}$ is a (positive) martingale with respect to the Q^{T_0}-forward measure, see Sect. 1.3.2. Thus, $(A_t)_{t \leq T_0}$ can be used as a numéraire to define the following change of measure

$$\left.\frac{dQ^{swap}}{dQ^{T_0}}\right|_{\mathscr{F}_t} = \frac{A_t}{p(t, T_0)} \frac{p(0, T_0)}{A_0} \tag{1.44}$$

Changing the measure to the swap measure in (1.43), the price of the swaption can be expressed as a price of a call option with strike R on the swap rate $R(T_0; T_0, T_n)$:

$$P^{Swn}(t; T_0, T_n, R) = A_t E^{Q^{swap}}\left\{\left(R(T_0; T_0, T_n) - R\right)^+ | \mathscr{F}_t\right\} \tag{1.45}$$

Remark 1.9 Recalling the definition of the OIS from Sect. 1.4.4, one could also consider an option with maturity $T = T_0$ to enter in an OIS as an underlying swap. Assume again that the nominal of the underlying OIS is set to $N = 1$. In this case, the price at time t of the corresponding swaption is given by (see 1.31)

$$P^{Swn}(t; T_0, T_n, R) = p(t, T_0) E^{Q^{T_0}} \left\{ \left(P^{OIS}(T_0; T_n, R) \right)^+ | \mathscr{F}_t \right\}$$

$$= p(t, T_0) E^{Q^{T_0}} \left\{ \left(1 - \sum_{k=1}^{n} c_k p(T_0, T_k) \right)^+ | \mathscr{F}_t \right\} \quad (1.46)$$

where $c_k = R\delta_k$, for $k = 1, \dots, n-1$, and $c_n = 1 + R\delta_n$, which can be recognized as a classical pre-crisis transformation of a swaption into a put option with strike 1 on a coupon bearing bond. Note that, as can be easily seen from Eqs. (1.27) and (1.19), such a pre-crisis transformation is no longer available in the post-crisis setup for Libor-indexed interest rate swaps. On the other hand, the expression for the OIS swaption price similar to (1.43)

$$P^{Swn}(t; T_0, T_n, R) = p(t, T_0) \sum_{k=1}^{n} \delta_k E^{Q^{T_0}} \left\{ p(T_0, T_k) \left(R^{OIS}(T_0; T_0, T_n) - R \right)^+ | \mathscr{F}_t \right\}$$

remains valid.

Remark 1.10 Similarly, a basis swaption is an option to enter into a basis swap. Considering a basis swap defined in Sect. 1.4.5 with nominal $N = 1$ as an underlying basis swap and $T_0 = T_0^1 = T_0^2$ as a maturity date of the option, the price of the basis swaption with strike basis spread S at time t is given by

$$P^{BSwn}(t; T_0, \mathscr{T}^1, \mathscr{T}^2, S)$$

$$= p(t, T_0) E^{Q^{T_0}} \left\{ \left(P^{BSw}(T_0; \mathscr{T}^1, \mathscr{T}^2, S) \right)^+ | \mathscr{F}_t \right\}$$

$$= p(t, T_0) E^{Q^{T_0}} \left\{ \left(\sum_{i=1}^{n_1} \delta_i^1 p(T_0, T_i^1) L(T_0; T_{i-1}^1, T_i^1) \right. \right.$$

$$\left. \left. - \sum_{j=1}^{n_2} \delta_j^2 p(T_0, T_j^2) (L(T_0; T_{j-1}^2, T_j^2) + S) \right)^+ | \mathscr{F}_t \right\}$$

$$= p(t, T_0) E^{Q^{T_0}} \left\{ \sum_{j=1}^{n_2} \delta_j^2 p(T_0, T_j^2) \left(S^{BSw}(T_0; \mathscr{T}^1, \mathscr{T}^2) - S \right)^+ | \mathscr{F}_t \right\} \quad (1.47)$$

by Eq. (1.40). Hence, we can define, analogously as for swaptions, a numéraire process

$$A_t^2 := \sum_{j=1}^{n_2} \delta_j^2 p(t, T_j^2)$$

and the corresponding (basis) swap measure

$$\frac{dQ^{swap,2}}{dQ^T}\bigg|_{\mathscr{F}_t} = \frac{A_t^2}{p(t,T_0)}\frac{p(0,T_0)}{A_0^2} \tag{1.48}$$

such that the price of the basis swaption can be expressed as a price of a call option on the basis swap spread $S^{BSw}(T_0; \mathscr{T}^1, \mathscr{T}^2)$:

$$P^{BSwn}(t; T_0, \mathscr{T}^1, \mathscr{T}^2, S) = A_t^2 E^{Q^{swap,2}}\left\{\left(S^{BSw}(T_0; \mathscr{T}^1, \mathscr{T}^2) - S\right)^+ \bigg| \mathscr{F}_t\right\} \tag{1.49}$$

Chapter 2
Short-Rate and Rational Pricing Kernel Models for Multiple Curves

In this chapter we shall consider mainly strict-sense (classical) short-rate models in view of constructing multiple curves. For this we shall base ourselves and partly extend previous work in this setting, in particular Kijima et al. (2009), Kenyon (2010), Filipović and Trolle (2013) and Morino and Runggaldier (2014). Because the pre-crisis rational pricing kernel models can also be considered as short-rate models in a wider sense, we shall furthermore present in this chapter some recent multiple curve extensions of these models based on Crépey et al. (2015b) and Nguyen and Seifried (2015).

The short rate will be given by the OIS short rate as it was introduced in Sect. 1.3.1 (see also the subsequent Sect. 1.3.2). We also recall here the relationship (1.13) that expresses the OIS bond prices in terms of the OIS short rate via an expectation under the martingale measure Q.

Starting from the strict-sense short-rate models, the setup which will be considered is the one of exponentially affine and exponentially quadratic models with several stochastic factors. These factors will be driven by Wiener processes, although the extension to jump-diffusions can be obtained in a relatively straightforward manner as will be briefly mentioned in Sect. 2.1.1. Such types of models have been developed and well studied in the context of interest rate modeling before the crisis and have been widely accepted and used because of their generality and flexibility coupled with analytical tractability. More precisely, by exploiting also results from the theory of affine processes we are able to obtain in this setting closed or semi-closed formulas for the prices of linear and optional interest rate derivatives.

When extending the strict-sense short-rate models we profit from the well-established pre-crisis exponentially affine and quadratic modeling approaches in order to develop suitable short-rate models of multiple curves and to price interest rate derivatives. In particular, in the case of linear derivatives, we shall directly compare the pre-crisis and the post-crisis derivative values and obtain "adjustment factors" allowing one to pass from one value to the other. For optional derivatives,

© The Author(s) 2015
Z. Grbac and W.J. Runggaldier, *Interest Rate Modeling: Post-Crisis Challenges and Approaches*, SpringerBriefs in Quantitative Finance, DOI 10.1007/978-3-319-25385-5_2

we shall show how to learn from the pre-crisis valuation methods to derive the post-crisis formulas for option prices. Therefore, the chapter will provide a full treatment of multiple curve pricing techniques applicable in the given modeling framework, with explicit formulas for each derivative type and model specification.

Moreover, we shall show that our exponentially affine model class includes the short-rate multi-curve models that have been introduced previously in the multi-curve literature (see Sect. 2.6). One peculiarity of our model class is also that we allow for correlation between the short rate and the spreads, which was not taken into account in most short-rate multi-curve models having appeared so far in the literature.

We shall consider the short rate r_t itself and a short-rate spread s_t to be added to the short rate from the outset. Since we may have one curve for each tenor Δ, we may have to consider a spread s_t^Δ for each tenor. Without loss of generality and in order to keep the presentation as simple as possible, we may consider just two different tenors Δ_1 and Δ_2. While r_t does not necessarily need to be restricted to take only positive values, this is however desirable for the spreads. In addition, also the difference between the spreads for two tenors Δ_1 and Δ_2 such that $\Delta_2 \geq \Delta_1$ should be positive as discussed in Sect. 1.2.1. Denoting then by s_t^1 and s_t^2 the spreads corresponding to the tenors Δ_1 and Δ_2 respectively, we let $\rho_t := s_t^2 - s_t^1$ and since, $s_t^2 = s_t^1 + \rho_t$, in what follows we shall consider as spread only s_t^1 and simply denote it by s_t. Summarizing we shall use the notation

$$r_t \;:\; \text{short rate}$$
$$s_t \;:\; \text{spread for tenor } \Delta_1$$
$$\rho_t \;:\; \text{difference between the spreads for tenors } \Delta_2 \text{ and } \Delta_1$$

To model the dynamics of (r_t, s_t, ρ_t) we shall introduce two classes of factor models for the short rate and the spreads. The two classes correspond to what for single-curve models may be called respectively *exponentially affine* (see Björk 2009, Brigo and Mercurio 2006, Filipović 2009) and *Gaussian, exponentially quadratic models* (see El Karoui et al. 1992, Pelsser 1997, Gombani and Runggaldier 2001, Leippold and Wu 2002, Chen et al. 2004, Gaspar 2004, Kijima et al. 2009). Note that in both classes positivity of factors can be obtained: in the exponentially affine class by choosing for example CIR processes as driving processes, where the positivity is ensured via the presence of the square-root in the defining SDE of the CIR processes, and in the exponentially quadratic class by applying the quadratic function to a Gaussian process. For sake of brevity of exposition, we shall concentrate mainly on the exponentially affine model class in the subsequent sections, referring for a more detailed treatment of the Gaussian, exponentially quadratic case to Grbac et al. (2015).

In the last part of the chapter we shall synthesize the main ideas concerning multi-curve extensions of rational pricing kernel models contained in Crépey et al. (2015b) and Nguyen and Seifried (2015). Rational pricing kernel models in the pre-crisis setting have been developed, among others, in Flesaker and Hughston (1996), Rogers (1997) and, more recently, in Filipović et al. (2014).

The chapter is organized as follows. The exponentially affine models are presented in Sect. 2.1, considering in particular the square-root and the Vasiček-type model classes. We emphasize that by Vasiček-type models we mean the Vasiček model itself, as well as the Hull-White extension of the model to time-dependent coefficients. For simplicity of presentation, the results will be given for the case of the Vasiček model, noting however that they can be extended in a straightforward manner to the Hull-White extension of this model. In Sect. 2.1.1 we describe more specifically the square-root and Vasiček-type exponentially affine models, Sect. 2.1.2 summarizes some technical preliminaries, and in Sect. 2.1.3 we shall derive explicit representations for the bond prices and the Libor rate in the exponentially affine model class. In Sect. 2.2 we summarize some relevant features of the Gaussian, exponentially quadratic model class. We shall then discuss pricing of linear and optional derivatives for the models of Sect. 2.1, following a same approach that applies to both model classes, namely the square-root and the Vasiček-type model class. In Sect. 2.3 the linear derivatives will be considered. We shall show that one can obtain an "adjustment factor" allowing one to pass from pre-crisis values to the corresponding post-crisis ones, in other words, from the one-curve to the multi-curve (two-curve) setting. Section 2.4 concerns the pricing of caps, while swaption pricing is studied in Sect. 2.5. Here it is also shown that the price of a swaption can be expressed as a linear combination of "caplet prices" with random strikes. In Sect. 2.6 we shall show more explicitly how other models from the literature, in particular the model of Filipović and Trolle (2013), can be obtained as special cases of our affine model class. Finally, Sect. 2.7 summarizes the main ideas on multiple curve rational pricing kernel models contained in Crépey et al. (2015b) and Nguyen and Seifried (2015).

2.1 Exponentially Affine Factor Models

2.1.1 The Factor Model and Properties

We shall consider a number of factor processes Ψ^i and model their dynamics under a martingale measure Q as mean-reverting square-root processes, namely

$$d\Psi_t^i = (a^i - b^i\Psi_t^i)dt + \sigma^i \sqrt{c^i\Psi_t^i + d^i}\, dw_t^i \qquad (2.1)$$

with w^i mutually independent Wiener processes so that also the factors are mutually independent. Furthermore, the coefficients are supposed to be positive, except for c^i and d^i that might also be zero (c^i equal to zero reduces (2.1) to a Vasiček model), and satisfying $a^i \geq \frac{c^i(\sigma^i)^2}{2}$ (such that in case $c^i > 0$ and $d^i = 0$ which yields a CIR model, the solution Ψ_t^i is a.s. positive).

We shall next model the short rate and the spreads in terms of the above factors. Although this has not been considered in the previous literature on multi-curve short-rate models (see, however, Morino and Runggaldier 2014), we shall also allow for

correlation between the short rate and the spreads. By analogy to the credit risk
setting, where the spread s_t corresponds to a default intensity, we shall model the
correlation by considering a common factor driving the short rate, as well as the
spreads (see e.g. Example 9.29 in McNeil et al. 2005). The other factors will be
considered as idiosyncratic factors and, always for the sake of keeping complexity
at a minimum, we shall have just one idiosyncratic factor for each of the quantities
r_t, s_t, ρ_t.

Finally, to allow for the possibility to derive below an "adjustment factor" between
pre- and post-crisis prices of linear derivatives, the common factor is supposed to satisfy a pure mean-reverting (Vasiček or Hull-White) model. In this way short rate and
short-rate spreads might take negative values, although only with small probability. A
possibility to prevent this from happening for the spreads is to model their dynamics
with an additive jump process taking only positive values, see also Sect. 2.6.2.

Considering, as discussed above, one common factor and one idiosyncratic factor
for the short rate and each of the spreads, we shall put

$$\begin{cases} r_t = \Psi_t^1 + \Psi_t^2 \\ s_t = \kappa^s \Psi_t^1 + \Psi_t^3 \\ \rho_t = \kappa^\rho \Psi_t^1 + \Psi_t^4 \end{cases} \tag{2.2}$$

with κ^s and κ^ρ expressing the intensity of the correlation and with Ψ_t^i satisfying, for
$i = 1, 2$,

$$\begin{cases} d\Psi_t^1 = (a^1 - b^1\Psi_t^1)dt + \sigma^1\,dw_t^1 \\ \\ d\Psi_t^2 = (a^2 - b^2\Psi_t^2)dt + \sigma^2\sqrt{c^2\Psi_t^2 + d^2}\,dw_t^2 \end{cases} \tag{2.3}$$

and, for $i = 3, 4$,

$$d\Psi_t^i = (a^i - b^i\Psi_t^i)dt + \sigma^i\sqrt{c^i\Psi_t^i + d^i}\,dw_t^i + \delta^i\mu^i(dt, dz) \tag{2.4}$$

with $\delta^i \geq 0$ ($\delta^i = 0$ reduces the dynamics to those of a mean-reverting square-root
process). In (2.4) $\mu^i(dt, dz)$ are jump measures where the compensators may be
chosen to be of the form

$$\nu^i(dt, dz) = \bar{\Psi}_t^i dt + \mu^i(dz) \tag{2.5}$$

with $\mu^i(\cdot)$ a probability measure over a given mark space, and with $\bar{\Psi}_t^i$ being additional
factors of the usual mean reverting square-root form

$$d\bar{\Psi}_t^i = (\bar{a}^i - \bar{b}^i\bar{\Psi}_t^i)dt + \bar{\sigma}^i\sqrt{\bar{\Psi}_t^i}\,d\bar{w}_t^i \quad i = 3, 4 \tag{2.6}$$

where $\bar{w}^i$ are independent Wiener processes, independent of those w^i in (2.3) and
(2.4). Notice that, with some assumptions on the jump measure, also jump diffusion
models as in (2.4) lead to an affine term structure (see Björk et al. 1997).

Remark 2.1 The coefficients a^i, $i = 1, 2, 3, 4$, in (2.3) and (2.4) have, for simplicity of presentation, been assumed to be constant parameters. They may more generally be taken to be time varying and the results below can be extended without additional conceptual difficulties to this more general case, for which only the computations become more involved. Note that taking time-varying coefficients in the first SDE in (2.3) yields the Hull-White extension of the Vasiček model. With time varying a^i in the factor dynamics one obtains time varying coefficients also in the dynamics, induced by (2.2), for the short rate and the spreads. Time varying coefficients have the advantage of allowing for a good fit to the initial term structure; on the other hand they lead to more complex calculations. Alternatively to time varying a^i one may, instead of (2.2), postulate for r_t a relation of the form

$$r_t = \Psi_t^1 + \Psi_t^2 + \phi_t$$

with ϕ_t a deterministic time function (*deterministic shift extension*). This leads to what is called a CIR2++ model, or, when Ψ_t^1, Ψ_t^2 both follow a Vasiček model (both are Gaussian processes) a G2++ model, see Brigo and Mercurio (2006). The "2" in CIR2++ and G2++ refers to the fact that there are two factors Ψ_t^1, Ψ_t^2, while the "++" refers to the deterministic shift. A generalization of the deterministic shift extension to the multi-curve setup has been developed in Grasselli and Miglietta (2014).

Remark 2.2 For what concerns the coefficients b^i that, as a^i, are modeled as constant parameters, below we shall occasionally end up with a time varying coefficient b^2 due to the transformation from the martingale measure Q to the forward measure $Q^{T+\Delta}$ in the CIR dynamics of Ψ_t^2. In particular, this happens in Sect. 2.4 where in (2.70) the parameters c^2 and λ^2 depend on t.

Finally, notice that the coefficients may even be adapted stochastic processes, which may be described by additional factors of the same form as the $\bar{\Psi}_t^i$ in (2.6); this is e.g. the case with the dynamics (30) and (33) in Filipović and Trolle (2013).

In order not to overburden the treatment, from now on we shall consider only affine factor models of the pure diffusion type and limit ourselves to the short rate and the single spread s_t defined as in (2.2) (the spread ρ_t can be treated by full analogy with s_t).

2.1.2 Technical Preliminaries

In this subsection we recall some known facts for the affine model class (2.1) that shall be used repeatedly below. We assume the processes to be given on a filtered probability space $(\Omega, \mathscr{F}, (\mathscr{F}_t)_{t \geq 0}, Q)$ with a martingale measure Q; we denote by E the expectation under Q.

The lemma below applies to the Hull-White/Vasiček version of the model class (2.1), where, as explained in the introduction, in order to keep the formulas simpler, we focus on the pure Vasiček model.

Lemma 2.1 *Consider the Vasiček version of the model class (2.1), namely, setting* $c^i = 0$, $d^i = 1$, *for a generic process* Ψ *we assume*

$$d\Psi_t = (a - b\Psi_t)dt + \sigma\, dw_t \tag{2.7}$$

For any $\gamma, K \in \mathbb{R}$ *we have the following*

$$E\left\{\exp\left[-\int_t^T \gamma\Psi_s ds - K\Psi_T\right]\bigg|\mathscr{F}_t\right\} = \exp\left[A(t, T) - B(t, T)\Psi_t\right] \tag{2.8}$$

where the coefficients satisfy

$$\begin{cases} B_t(t, T) - bB(t, T) + \gamma = 0 & , \; B(T, T) = K \\ A_t(t, T) = aB(t, T) - \frac{\sigma^2}{2}B^2(t, T) & , \; A(T, T) = 0 \end{cases} \tag{2.9}$$

and are given by

$$\begin{cases} B(t, T) = -\frac{\gamma}{b}\left[\left(-\frac{bK}{\gamma} + 1\right)e^{-b(T-t)} - 1\right] \\ \qquad = Ke^{-b(T-t)} - \frac{\gamma}{b}\left(e^{-b(T-t)} - 1\right) \\ A(t, T) = -a\int_t^T B(s, T)ds + \frac{\sigma^2}{2}\int_t^T (B(s, T))^2 ds \end{cases} \tag{2.10}$$

In the last expression the integrals are easy to compute, but lead to lengthy expressions. The proof can be found in most textbooks treating the affine term structure.

Lemma 2.2 *Consider the CIR version of the model class (2.1), namely, setting* $c^i = 1$, $d^i = 0$, *for a generic process* Ψ *we assume*

$$d\Psi_t = (a - b\Psi_t)dt + \sigma\sqrt{\Psi_t}\, dw_t \tag{2.11}$$

For any $T > 0$ *define the set*

$$\mathscr{I}_T := \left\{u \in \mathbb{R} : E\left\{e^{u\Psi_T}\right\} < \infty\right\} \tag{2.12}$$

i.e. the set of $u \in \mathbb{R}$ *for which the moment generating function of* Ψ_T *is well-defined. Then for any* $\gamma > 0$ *and any* $K \in \mathbb{R}$ *such that* $-K \in \mathscr{I}_T$ *we have the following*

$$E\left\{\exp\left[-\int_t^T \gamma\Psi_s ds - K\Psi_T\right]\bigg|\mathscr{F}_t\right\} = \exp\left[A(t, T) - B(t, T)\Psi_t\right] \tag{2.13}$$

where the coefficients satisfy

$$\begin{cases} B_t(t, T) - bB(t, T) - \frac{\sigma^2}{2}B^2(t, T) + \gamma = 0 & , \; B(T, T) = K \\ A_t(t, T) = aB(t, T) & , \; A(T, T) = 0 \end{cases} \tag{2.14}$$

and are given by

$$
\begin{cases}
B(t,T) = \dfrac{K\left(h+b+e^{h(T-t)}(h-b)\right)+2\gamma\left(e^{h(T-t)}-1\right)}{K\sigma^2\left(e^{h(T-t)}-1\right)+h-b+e^{h(T-t)}(h+b)}, & \text{where } h := \sqrt{b^2+2\gamma\sigma^2} \\[4mm]
A(t,T) = -a\int_t^T B(s,T)ds = \dfrac{2a}{\sigma^2}\log\left(\dfrac{2he^{\frac{(T-t)(h+b)}{2}}}{K\sigma^2\left(e^{h(T-t)}-1\right)+h-b+e^{h(T-t)}(h+b)}\right)
\end{cases}
\tag{2.15}
$$

The proof can also be found in many textbooks treating the affine term structure. The explicit expressions reported here come from Lamberton and Lapeyre (2007). Note that the result therein is provided for $K > 0$ (which obviously satisfies $-K \in \mathscr{I}_T$), but it can be extended in a straightforward manner also to the case $K < 0$ satisfying $-K \in \mathscr{I}_T$.

Remark 2.3 Although $A(t,T)$ and $B(t,T)$ in Lemmas 2.1 and 2.2 depend on the constants γ and K, we do not make this dependence explicit in order not overburden the notation. The specific values of γ and K to be used in the various applications of the two lemmas in the sequel will become clear from the context.

Remark 2.4 The lemma above will be applied several times in this chapter, also under different forward measures. If the expectation in (2.12) is given with respect to the forward measure Q^S, for any generic date $S > 0$, then we set:

$$
\mathscr{I}_T^S := \left\{u \in \mathbb{R} : E^{Q^S}\left\{e^{u\Psi_T}\right\} < \infty\right\}
\tag{2.16}
$$

In view of the next lemma recall from (1.14) that the density process for a change from Q to the forward measure $Q^{T+\Delta}$ is

$$
\mathscr{L}_t = \left.\frac{dQ^{T+\Delta}}{dQ}\right|_{\mathscr{F}_t} = \frac{p(t,T+\Delta)}{p(0,T+\Delta)}\frac{1}{B_t}
\tag{2.17}
$$

leading to

$$
\begin{aligned}
d\mathscr{L}_t &= \frac{1}{p(0,T+\Delta)}\left(p(t,T+\Delta)d(B_t^{-1}) + \frac{dp(t,T+\Delta)}{B_t}\right) \\
&= \frac{1}{p(0,T+\Delta)}\left(-\frac{p(t,T+\Delta)B_t r_t dt}{B_t^2} + \frac{dp(t,T+\Delta)}{B_t}\right)
\end{aligned}
\tag{2.18}
$$

Lemma 2.3 *Consider three factor processes* Ψ^1, Ψ^2, Ψ^3 *satisfying, under Q, the following system related to (2.3),*

$$
\begin{cases}
d\Psi_t^1 = (a^1 - b^1\Psi_t^1)dt + \sigma^1\,dw_t^1 \\[2mm]
d\Psi_t^i = (a^i - b^i\Psi_t^i)dt + \sigma^i\sqrt{c^i\Psi_t^i + d^i}\,dw_t^i, \quad i = 2,3
\end{cases}
\tag{2.19}
$$

Then under the $(T + \Delta)$-forward measure the same processes satisfy

$$
\begin{cases}
d\Psi_t^1 = \left(a^1 - (\sigma^1)^2 B^1(t, T + \Delta) - b^1 \Psi_t^1\right) dt + \sigma^1 \, dw_t^{1,T+\Delta} \\
d\Psi_t^2 = \left(a^2 - d^2(\sigma^2)^2 B^2(t, T + \Delta) - \Psi_t^2(b^2 + c^2(\sigma^2)^2 B^2(t, T + \Delta))\right) dt \\
\qquad + \sigma^2 \sqrt{c^2 \Psi_t^2 + d^2} \, dw_t^{2,T+\Delta} \\
d\Psi_t^3 = (a^3 - b^3 \Psi_t^3) dt + \sigma^3 \sqrt{c^3 \Psi_t^3 + d^3} \, dw_t^{3,T+\Delta}
\end{cases}
$$

$$(2.20)$$

where $w^{i,T+\Delta}$, $i = 1, 2, 3$, are independent $Q^{T+\Delta}$-Wiener processes and $B^1(t, T)$ corresponds to the $B(t, T)$ in Lemma 2.1, while $B^2(t, T)$ corresponds to the $B(t, T)$ in Lemma 2.2, in both cases with $\gamma = 1$ and $K = 0$.

Proof For the short rate r_t expressed according to (2.2) in terms of the factors Ψ_t^i, $i = 1, 2$, that satisfy the Q-dynamics (2.19), we have from Lemmas 2.1 and 2.2 with $\gamma = 1$ and $K = 0$, as well as from (1.13), that the OIS bond prices can be expressed as

$$
p(t, T) = E^Q \left\{ \exp\left[- \int_t^T (\Psi_u^1 + \Psi_u^2) \, du \right] \Big| \mathscr{F}_t \right\}
$$
$$
= \exp\left[A(t, T) - B^1(t, T) \Psi_t^1 - B^2(t, T) \Psi_t^2 \right] \qquad (2.21)
$$

where $A(t, T) := A^1(t, T) + A^2(t, T)$, for all $T \geq 0$ and $t \leq T$. Under the measure Q the dynamics of $p(t, T + \Delta)$ is then

$$
dp(t, T + \Delta) = p(t, T + \Delta)\Big(r_t \, dt - \sigma^1 B^1(t, T + \Delta) \, dw_t^1
$$
$$
- \sigma^2 B^2(t, T + \Delta) \sqrt{c^2 \Psi_t^2 + d^2} \, dw_t^2 \Big) \qquad (2.22)
$$

implying that

$$
d\mathscr{L}_t = \mathscr{L}_t \Big(-\sigma^1 B^1(t, T + \Delta) \, dw_t^1 - \sigma^2 B^2(t, T + \Delta) \sqrt{c^2 \Psi_t^2 + d^2} \, dw_t^2 \Big) \quad (2.23)
$$

Let then

$$
\begin{cases}
dw_t^{1,T+\Delta} = dw_t^1 + \sigma^1 B^1(t, T + \Delta) \, dt \\
dw_t^{2,T+\Delta} = dw_t^2 + \sigma^2 B^2(t, T + \Delta) \sqrt{c^2 \Psi_t^2 + d^2} \, dt \\
dw_t^{3,T+\Delta} = dw_t^3
\end{cases} \qquad (2.24)
$$

By the multidimensional Girsanov theorem the processes in (2.24) are Wiener processes under the measure $Q^{T+\Delta}$ thus leading to the statement of the lemma. $\square$

Lemma 2.4 *On a given filtered probability space with filtration $(\mathscr{F}_t)_{t \geq 0}$ and an $(\mathscr{F}_t)$-Wiener process w consider the equation*

$$dX_t = (a_t X_t + b_t)\,dt + \sigma_t dw_t \tag{2.25}$$

with a_t, b_t, σ_t bounded and adapted processes. Then the equation has a unique strong solution given by

$$X_t = \Phi_t\left(X_0 + \int_0^t \Phi_s^{-1} b_s ds + \int_0^t \Phi_s^{-1}\sigma_s dw_s\right),\ \ t \geq 0 \tag{2.26}$$

where Φ_t is the fundamental solution satisfying

$$\frac{d}{dt}\Phi_t = a_t \Phi_t,\ t \geq 0,\quad \Phi_0 = 1$$

The proof can be found in most textbooks on stochastic analysis. In the next lemma we give explicit distributions for the three factors $\Psi_t^j, j = 1, 2, 3$. This is possible in two cases: when $c^i = 1, d^i = 0$ (CIR case), or when $c^i = 0, d^i = 1$ (Vasiček case), for $i = 2, 3$.

Lemma 2.5 *Denote by $N(\alpha, \beta)$ a Gaussian distribution with mean α and variance β. Denote, furthermore, by $\chi^2(v, \lambda)$ a non-central χ^2-distribution with v degrees of freedom and non-centrality parameter λ.*

The factor process Ψ^1, with dynamics given by (2.19) under Q, has the following distribution under $Q^{T+\Delta}$:

$$\Psi_t^1 \sim N(\alpha_t^1, \beta_t^1) \tag{2.27}$$

with the parameters given by

$$
\begin{cases}
\alpha_t^1 &= e^{-b^1 t}\left(\Psi_0^1 + \left(\dfrac{a^1}{b^1} - \dfrac{(\sigma^1)^2}{(b^1)^2}\right)(e^{b^1 t} - 1) + \dfrac{(\sigma^1)^2}{2(b^1)^2}e^{-b^1(T+\Delta)}(e^{2b^1 t} - 1)\right) \\
\beta_t^1 &= \dfrac{(\sigma^1)^2}{2(b^1)}e^{-2b^1 t}(e^{2b^1 t} - 1)
\end{cases}
\tag{2.28}
$$

For $c^i = 1, d^i = 0, i = 2, 3$, the two factor processes Ψ^2, Ψ^3 satisfying, under Q, the system (2.19), under $Q^{T+\Delta}$ have the following distributions

$$\Psi_t^2 \sim \frac{\chi^2(v^2, \lambda_t^2)}{c_t^2},\quad \Psi_t^3 \sim \frac{\chi^2(v^3, \lambda_t^3)}{c_t^3} \tag{2.29}$$

with the parameters given by

$$
\begin{cases}
c_t^2 &= \dfrac{4(b^2 + (\sigma^2)^2 B^2(t, T + \Delta))}{(\sigma^2)^2 (1 - e^{-t(b^2 + (\sigma^2)^2 B^2(t, T+\Delta))})} \\[4pt]
v^2 &= \dfrac{4a^2}{(\sigma^2)^2} \\[4pt]
\lambda_t^2 &= c_t^2 \Psi_0^2 e^{-t(b^2 + (\sigma^2)^2 B^2(t, T+\Delta))} \\[4pt]
v^3 &= \dfrac{4a^3}{(\sigma^3)^2} \\[4pt]
c_t^3 &= \dfrac{4b^3}{(\sigma^3)^2 (1 - e^{-b^3 t})} \\[4pt]
\lambda_t^3 &= c_t^3 \Psi_0^3 e^{-b^3 t}
\end{cases}
\tag{2.30}
$$

where $B^2(t, T)$ corresponds to the $B(t, T)$ in Lemma 2.2.

For $c^i = 0$, $d^i = 1$, $i = 2, 3$, the two factor processes Ψ^2, Ψ^3 satisfying, under Q, the system (2.19), under $Q^{T+\Delta}$ have the following distributions

$$
\Psi_t^2 \sim N(\alpha_t^2, \beta_t^2), \quad \Psi_t^3 \sim N(\alpha_t^3, \beta_t^3)
\tag{2.31}
$$

with the parameters given by

$$
\begin{cases}
\alpha_t^2 &= e^{-b^2 t}\left(\Psi_0^2 + \left(\dfrac{a^2}{b^2} - \dfrac{(\sigma^2)^2}{(b^2)^2}\right)(e^{b^2 t} - 1) + \dfrac{(\sigma^2)^2}{2(b^2)^2} e^{-b^2(T+\Delta)}(e^{2b^2 t} - 1)\right) \\[6pt]
\beta_t^2 &= \dfrac{(\sigma^2)^2}{2(b^2)} e^{-2b^2 t}(e^{2b^2 t} - 1) \\[6pt]
\alpha_t^3 &= e^{-b^3 t}\left(\Psi_0^3 + \dfrac{a^3}{b^3}(e^{b^3 t} - 1)\right) \\[6pt]
\beta_t^3 &= \dfrac{(\sigma^3)^2}{2(b^3)} e^{-2b^3 t}(e^{2b^3 t} - 1)
\end{cases}
\tag{2.32}
$$

Proof For the factor process Ψ_t^1 the proof follows immediately from (2.20) using Lemma 2.4, as well as for the processes Ψ_t^i, $i = 2, 3$ in case $c^i = 0$, $d^i = 1$. For the processes Ψ_t^i, $i = 2, 3$ in case $c^i = 1$, $d^i = 0$, we refer to Brigo and Mercurio (2006), Sect. 3.2.3, always using the dynamics in (2.20). □

Remark 2.5 In defining the dynamics of the factor processes Ψ_t^i in (2.1) we followed the traditional "martingale modeling" whereby the dynamics are assigned in the standard martingale measure Q with numéraire B_t. This leads to the standard way of defining the OIS bond prices $p(t, T)$ according to (1.13) and, analogously, $\bar{p}(t, T)$ in (2.35). On the other hand, for the pricing of the interest rate derivatives we shall use the forward measure that implies the more complex dynamics for the factors as specified in (2.20) and this in turn induces more complex pricing formulas. One might then expect that a direct modeling of the factor dynamics under the forward measure might be computationally more advantageous. This is indeed true as long as one has to deal with only one forward measure $Q^{T+\Delta}$ as it would occur in the pricing of FRAs and single caplets. However, for the majority of the derivatives, namely full caps and floors, as well as swaps and swaptions, one has to consider different forward measures, one for each interval of the tenor structure so that one cannot avoid the more complex dynamics resulting from a measure transformation.

2.1.3 Explicit Representation of the Libor Rate

As discussed in detail in Sect. 1.4.1, in the post-crisis fixed-income markets multiple yield curves have appeared. In the following description we may without loss of generality consider just a single generic tenor Δ and set $S = T + \Delta$. Recalling that the Libor rates $L(T; T, T + \Delta)$ are determined by the Libor panel that takes into account various factors such as credit risk, liquidity, etc., in the post-crisis framework these rates have to be considered as risky and, as it is documented also empirically, one has that in general

$$L(T; T, T + \Delta) \neq F(T; T, T + \Delta) = \frac{1}{\Delta}\left(\frac{1}{p(T, T + \Delta)} - 1\right)$$

thus leading to a *Libor-OIS spread*, cf. Eq. (1.33).

In this chapter, we shall take the approach outlined in Remark 1.4. More precisely, we keep the classical formal relationship between discretely compounding forward rates and bond prices also for the Libor rates, but replace the bond prices $p(t, T)$ by fictitious ones $\bar{p}(t, T)$ that are supposed to be affected by the same factors as the Libor rates. Recall that the fictitious bonds are not traded assets, but one still assumes $\bar{p}(T, T) = 1$. Actually, since discretely compounding forward rates are usually given by FRA rates where at the maturity of the FRA a fixed rate payment is made in exchange of the Libor rate $L(T; T, T + \Delta)$ payment, we postulate the classical relationship only at the inception time $t = T$. Since the various interest derivatives concern the spot Libor rates, it is thus sufficient to define the Libor rates in our context as being given by

$$L(T; T, T + \Delta) = \frac{1}{\Delta}\left(\frac{1}{\bar{p}(T, T + \Delta)} - 1\right) \tag{2.33}$$

This implies that we need explicit expressions for $\bar{p}(T, T + \Delta)$ and, more generally, for $\bar{p}(t, T)$. To this effect recall that, in a pure credit risk setting where τ denotes the default time of the counterparty, the price $p^d(t, T)$ of a defaultable bond with zero recovery can be expressed as

$$p^d(t, T) = E^Q\left\{\exp\left[-\int_t^T r_s ds\right] 1_{\{\tau > T\}} \Big| \mathcal{G}_t\right\} \tag{2.34}$$

where $\mathcal{G}_t = \mathcal{F}_t \vee \mathcal{H}_t$, with $(\mathcal{H}_t)_{0 \leq t \leq T^*}$ denoting the filtration generated by the default indicator process $1_{\{\tau \leq t\}}$. Using the intensity approach to credit risk, $p^d(t, T)$ in (2.34) can be given the following form

$$p^d(t, T) = 1_{\{\tau > t\}} \, \bar{p}(t, T)$$

where

$$\bar{p}(t, T) = E^Q \left\{ \exp \left[-\int_t^T (r_u + s_u)du \right] \Big| \mathcal{F}_t \right\} \qquad (2.35)$$

with s_t representing the hazard rate (default intensity). In the present context we assume the relation (2.35) also for our fictitious bonds thereby considering s_t to represent more generically a "short-rate spread" that incorporates not only credit risk, but also the various other risks in the interbank sector that affect the Libor rates. Since here we consider a single generic tenor Δ, we also consider only one spread s_t. Below we shall obtain explicit expressions for $\bar{p}(t, T)$ and this will then allow us to obtain explicit expressions also for derivative prices and adjustment factors.

Corresponding to (2.2) and basing ourselves on Morino and Runggaldier (2014), we shall consider the model

$$\begin{cases} r_t = \Psi_t^1 + \Psi_t^2 \\ s_t = \kappa \, \Psi_t^1 + \Psi_t^3 \end{cases} \qquad (2.36)$$

where κ corresponds to κ^s in (2.2) and Ψ_t^1, Ψ_t^2 satisfy (2.3). To avoid more complicated notation, we shall without loss of generality assume in this section, as well as in the following analogous ones, that Ψ_t^3 also satisfies just a mean-reverting square-root model of the same form as Ψ_t^2 instead of a jump-diffusion model as defined in (2.4).

In the sequel we shall study two affine model specifications: the common factor Ψ^1 is always of the mean-reverting Vasiček type, and the factors Ψ^2 and Ψ^3 are given in the first specification as CIR processes ($d^i = 0$) and in the second one as mean-reverting Vasiček-type processes ($c^i = 0$). The first specification corresponds thus to a factor model with the three factors Ψ_t^1, Ψ_t^2, Ψ_t^3 that, under the measure Q, are assumed to satisfy

$$\begin{cases} d\Psi_t^1 = (a^1 - b^1 \Psi_t^1)dt + \sigma^1 \, dw_t^1 \\ d\Psi_t^2 = (a^2 - b^2 \Psi_t^2)dt + \sigma^2 \sqrt{\Psi_t^2} \, dw_t^2 \\ d\Psi_t^3 = (a^3 - b^3 \Psi_t^3)dt + \sigma^3 \sqrt{\Psi_t^3} \, dw_t^3 \end{cases} \qquad (2.37)$$

and the second one to the three factors Ψ_t^1, Ψ_t^2, Ψ_t^3 given, under the measure Q, by

$$\begin{cases} d\Psi_t^1 = (a^1 - b^1 \Psi_t^1)dt + \sigma^1 \, dw_t^1 \\ d\Psi_t^2 = (a^2 - b^2 \Psi_t^2)dt + \sigma^2 \, dw_t^2 \\ d\Psi_t^3 = (a^3 - b^3 \Psi_t^3)dt + \sigma^3 \, dw_t^3 \end{cases} \qquad (2.38)$$

We first recall the expression for the OIS bond prices that is given by (see (2.21))

$$p(t, T) = \exp\left[A(t, T) - B^1(t, T)\Psi_t^1 - B^2(t, T)\Psi_t^2\right] \tag{2.39}$$

From Lemma 2.1 we obtain in particular

$$B^1(t, T) = -\frac{1}{b^1}\left(e^{-b^1(T-t)} - 1\right) \tag{2.40}$$

On the other hand, for the fictitious bond prices $\bar{p}(t, T)$ we have according to (2.35),

$$
\begin{aligned}
\bar{p}(t, T) &= E^Q\left\{\exp\left[-\int_t^T (r_u + s_u)du\right]\Big|\mathscr{F}_t\right\} \\
&= E^Q\left\{\exp\left[-\int_t^T ((\kappa + 1)\Psi_u^1 + \Psi_u^2 + \Psi_u^3)du\right]\Big|\mathscr{F}_t\right\} \\
&= \exp\left[\bar{A}(t, T) - \bar{B}^1(t, T)\Psi_t^1 - \bar{B}^2(t, T)\Psi_t^2 - \bar{B}^3(t, T)\Psi_t^3\right]
\end{aligned} \tag{2.41}
$$

where, analogously to $A(t, T)$, $B^i(t, T)$, the functions $\bar{A}(t, T)$, $\bar{B}^1(t, T)$, $\bar{B}^2(t, T)$ and $\bar{B}^3(t, T)$ are determined according to Lemmas 2.1 and 2.2 with $\bar{A}(t, T) = \bar{A}^1(t, T) + \bar{A}^2(t, T) + \bar{A}^3(t, T)$. In particular, for $\bar{B}^1(t, T)$ we have

$$\bar{B}^1(t, T) = -\frac{\kappa + 1}{b^1}\left(e^{-b^1(T-t)} - 1\right) = (1 + \kappa)\, B^1(t, T) \tag{2.42}$$

while $\bar{B}^2(t, T) = B^2(t, T)$. The linear relationship between $B^1(t, T)$ and $\bar{B}^1(t, T)$ will be important below to derive an "adjustment factor". For this reason in the model class (2.37) we postulated a Vašiček-type model for the common factor Ψ_t^1, while the other, idiosyncratic, factors are assumed to follow a mean-reverting CIR model.

Summarizing, from (2.39), (2.41) and taking (2.42) into account, we have the following

Proposition 2.1 *Assume that the OIS short rate r and the spread s are given by (2.36) with the factor processes Ψ_t^i, $i = 1, 2, 3$, evolving according to either (2.37) or (2.38) under the standard martingale measure Q. The time-t price of the OIS bond $p(t, T)$, as given in (1.13), is obtained as*

$$p(t, T) = \exp\left[A(t, T) - B^1(t, T)\Psi_t^1 - B^2(t, T)\Psi_t^2\right], \tag{2.43}$$

and the time-t price of the fictitious bond $\bar{p}(t, T)$, as defined in (2.35), by

$$
\begin{aligned}
\bar{p}(t, T) &= \exp\left[\bar{A}(t, T) - \bar{B}^1(t, T)\Psi_t^1 - B^2(t, T)\Psi_t^2\right. \\
&\qquad\left. - \bar{B}^3(t, T)\Psi_t^3 - \kappa B^1(t, T)\Psi_t^1\right] \\
&= p(t, T)\, \exp\left[\tilde{A}(t, T) - \kappa B^1(t, T)\Psi_t^1 - \bar{B}^3(t, T)\Psi_t^3\right]
\end{aligned} \tag{2.44}
$$

with $\tilde{A}(t, T) := \bar{A}(t, T) - A(t, T)$, where $A(t, T)$ and $\bar{A}(t, T)$, as well as $B^2(t, T)$ and $\bar{B}^3(t, T)$, are expressed according to Lemmas 2.1 and 2.2 and where $B^1(t, T)$ is explicitly given by (2.40). Moreover, for the Libor rate $L(T; T, T + \Delta)$ in (2.33) we have

$$L(T; T, T + \Delta) = \frac{1}{\Delta} \left(\frac{\exp\left[-\tilde{A}(T,T+\Delta) + \kappa B^1(T,T+\Delta)\Psi_T^1 + B^3(T,T+\Delta)\Psi_T^3 \right]}{p(T,T+\Delta)} - 1 \right) \quad (2.45)$$

Notice that (2.44) allows one to express $\bar{p}(t, T)$ in terms of $p(t, T)$. Based on this relationship in Sect. 2.3.2 we shall derive an adjustment factor allowing to pass from pre-crisis quantities to the corresponding post-crisis quantities.

2.2   Gaussian, Exponentially Quadratic Models

In the exponentially affine model class (2.37) we had postulated a CIR dynamics for all the factors, except for the common factor Ψ_t^1 that satisfies a mean-reverting Vasiček-type model. This guarantees sufficient tractability and, except for a small probability, positivity of rates and spreads. The positivity implied by the CIR dynamics comes at the expense of a chi-square distribution for the corresponding factors (see (2.29), (2.30)), which is computationally more cumbersome to deal with than a Gaussian distribution. An alternative that has similar advantages but leads to Gaussianity for all the factors is to assume a Vasiček or Hull-White dynamics for all the factors but, instead of (2.2), postulating that

$$\begin{cases} r_t = \Psi_t^1 + (\Psi_t^2)^2 \\ s_t = \kappa^s \Psi_t^1 + (\Psi_t^3)^2 \\ \rho_t = \kappa^\rho \Psi_t^1 + (\Psi_t^4)^2 \end{cases} \quad (2.46)$$

Limiting ourselves for descriptive purposes to a single tenor Δ and thus to spot Libor rates of the form $L(T; T, T + \Delta)$ as done in Sect. 2.1.3, we may analogously to (2.36) consider the simpler system

$$\begin{cases} r_t = \Psi_t^1 + (\Psi_t^2)^2 \\ s_t = \kappa \Psi_t^1 + (\Psi_t^3)^2 \end{cases} \quad (2.47)$$

For the three factors $\Psi_t^1, \Psi_t^2, \Psi_t^3$ we shall assume that, under the measure Q, they satisfy a Gaussian model of the form (2.38), namely the following dynamics that for simplicity we take as zero mean-reverting

$$d\Psi_t^i = -b^i \Psi_t^i dt + \sigma^i dw_t^i, \quad i = 1, 2, 3 \quad (2.48)$$

Using results from Gombani and Runggaldier (2001) it can be shown that, corresponding to (2.39) and (2.41), the OIS bond prices $p(t, T)$ and the fictitious ones $\bar{p}(t, T)$ satisfy the following relations

$$p(t, T) = E^Q \left\{ e^{-\int_t^T r_u du} \mid \mathscr{F}_t \right\} = E^Q \left\{ e^{-\int_t^T (\Psi_u^1 + (\Psi_u^2)^2) du} \mid \mathscr{F}_t \right\}$$
$$= \exp \left[A(t, T) - \sum_{i=1}^3 B^i(t, T) \Psi_t^i - \sum_{i,j=1}^3 C^{ij}(t, T) \Psi_t^i \Psi_t^j \right]$$

and

$$\bar{p}(t, T) = E^Q \left\{ e^{-\int_t^T (r_u + s_u) du} \mid \mathscr{F}_t \right\} = E^Q \left\{ e^{-\int_t^T ((1+\kappa)\Psi_u^1 + (\Psi_u^2)^2 + (\Psi_u^3)^2) du} \mid \mathscr{F}_t \right\}$$
$$= \exp \left[\bar{A}(t, T) - \sum_{i=1}^3 \bar{B}^i(t, T) \Psi_t^i - \sum_{i,j=1}^3 \bar{C}^{ij}(t, T) \Psi_t^i \Psi_t^j \right]$$

respectively, where the coefficients satisfy suitable first order differential equations that, for the second order coefficients $C(t, T)$, are given by a matrix Riccati equation, while the remaining coefficients satisfy ordinary first order differential equations whereby, as in the exponentially affine case, $B(t, T)$ is determined on the basis of the values of $C(t, T)$, and $A(t, T)$ on those of $C(t, T)$ as well as $B(t, T)$. Analogously for $\bar{C}(t, T), \bar{B}(t, T), \bar{A}(t, T)$. It thus follows that, with the dynamics (2.48) and the relations (2.47), one ends up with an exponentially quadratic term structure where all the factors are Gaussian. This guarantees good computational tractability and, as for the affine model class of Sect. 2.1, positivity of short rates and spreads except for a small probability (see Kijima and Muromachi (2015) for an alternative representation guaranteeing positivity of s_t).

Since the results that we shall derive below for the pricing of linear and nonlinear derivatives for the case of the exponentially affine class can be extended to the Gaussian exponentially quadratic class without additional conceptual difficulties, we shall not report them here. For this, as well as other details concerning the subject of this section, we refer to Grbac et al. (2015).

2.3 Pricing of FRAs and Other Linear Derivatives

We start by recalling the notion of an FRA as given in Definition 1.3. As we have seen in Sect. 1.4, FRAs are prototypes of other linear derivatives such as interest rate swaps and so we shall concentrate here mainly on the pricing of an FRA.

Consider an FRA with the inception date T, the maturity $T + \Delta$, the fixed rate R and the notional N. Having just a single tenor, the no-arbitrage price of the FRA at $t < T$ is then given, as in (1.17), by

$$
\begin{aligned}
P^{FRA}(t; T, T + \Delta, R, N) &= N\Delta p(t, T + \Delta) E^{T+\Delta}\left\{ L(T; T, T + \Delta) - R \mid \mathscr{F}_t \right\} \\
&= N p(t, T + \Delta) E^{T+\Delta}\left\{ \frac{1}{\bar{p}(T, T + \Delta)} - (1 + \Delta R) \Big| \mathscr{F}_t \right\}
\end{aligned}
\tag{2.49}
$$

having used the underlying relationship between Libor rates and the fictitious bond prices as established in (2.33) and where for the pricing we have used the $(T + \Delta)$-forward measure (see Sect. 1.3.2). Notice also that the simultaneous presence of $p(t, T + \Delta)$ and $\bar{p}(t, T + \Delta)$ does not allow for a convenient reduction of the formula to a simpler form as in the one-curve setup. From (2.49) we see that the crucial quantity to compute in order to obtain the value $P^{FRA}(t; T, T + \Delta, R, N)$ is

$$
\bar{\nu}_{t,T} := E^{T+\Delta}\left\{ \frac{1}{\bar{p}(T, T + \Delta)} \Big| \mathscr{F}_t \right\}
\tag{2.50}
$$

and it is crucial also for all other linear derivatives. It follows that the fixed rate to make the FRA a fair contract at t, namely such that $P^{FRA}(t; T, T + \Delta, R, N) = 0$, is then

$$
\bar{R}_t = \frac{1}{\Delta}(\bar{\nu}_{t,T} - 1)
\tag{2.51}
$$

Notice also that this is exactly the forward Libor rate as defined in Definition 1.2, hence we have

$$
L(t; T, T + \Delta) = \frac{1}{\Delta}(\bar{\nu}_{t,T} - 1)
\tag{2.52}
$$

2.3.1 Computation of FRA Prices and FRA Rates

To compute now the value of $\bar{\nu}_{t,T}$ in (2.50), notice that the expectation there is under the $(T + \Delta)$-forward measure, while the dynamics of the factors $\psi_t^1, \psi_t^2, \psi_t^3$ that according to (2.41) drive the fictitious bond price process $\bar{p}(t, T)$ are defined under the standard martingale measure Q. We therefore perform a change from Q to the forward measure $Q^{T+\Delta}$ recalling from (2.17) that the corresponding density process is $\mathscr{L}_t = \frac{p(t, T+\Delta)}{p(0, T+\Delta) B_t}$. We can thus write

$$
\begin{aligned}
\bar{\nu}_{t,T} &= E^{T+\Delta}\left\{ \frac{1}{\bar{p}(T, T + \Delta)} \Big| \mathscr{F}_t \right\} = \mathscr{L}_t^{-1} E^Q\left\{ \frac{\mathscr{L}_T}{\bar{p}(T, T + \Delta)} \Big| \mathscr{F}_t \right\} \\
&= \frac{1}{p(t, T + \Delta)} E^Q\left\{ \exp\left[-\int_t^T r_u du \right] \frac{p(T, T + \Delta)}{\bar{p}(T, T + \Delta)} \Big| \mathscr{F}_t \right\}
\end{aligned}
\tag{2.53}
$$

Recalling the relationship (2.44) between $p(t, T)$ and $\bar{p}(t, T)$, one may write

$$\frac{p(T, T + \Delta)}{\bar{p}(T, T + \Delta)} = \exp\left[-\tilde{A}(T, T + \Delta) + \kappa B^1(T, T + \Delta)\Psi_T^1 + \bar{B}^3(T, T + \Delta)\Psi_T^3\right]$$

(2.54)

With this last expression Eq. (2.53) becomes

$$\begin{aligned}
\bar{\nu}_{t,T} &= \frac{1}{p(t, T + \Delta)} E^Q \left\{ \exp\left[- \int_t^T r_u du \right] \right. \\
&\qquad \cdot \exp\left[- \tilde{A}(T, T + \Delta) + \kappa B^1(T, T + \Delta)\Psi_T^1 \right. \\
&\qquad \left. \left. + \bar{B}^3(T, T + \Delta)\Psi_T^3 \right] \mid \mathscr{F}_t \right\} \\
&= \frac{1}{p(t, T + \Delta)} \exp\left[-\tilde{A}(T, T + \Delta)\right] E^Q \left\{ e^{\bar{B}^3(T,T+\Delta)\Psi_T^3} \mid \mathscr{F}_t \right\} \\
&\qquad \cdot E^Q \left\{ e^{-\int_t^T (\Psi_u^1 + \Psi_u^2)du} e^{\kappa B^1(T,T+\Delta)\Psi_T^1} \mid \mathscr{F}_t \right\}
\end{aligned}$$

(2.55)

Given the independence of the factors and the affine structure of their dynamics under Q, namely (2.37) and (2.38), the two expectations in the second equality in (2.55) can easily be calculated explicitly, see Lemmas 2.1 and 2.2, as exponentially affine functions of Ψ_t^1, Ψ_t^2 and Ψ_t^3. In particular, for the model (2.37) we have to assume that $\bar{B}^3(T, T + \Delta) \in \mathscr{I}_T$, so that Lemma 2.2 can be applied. From the expression for $\bar{\nu}_{t,T}$ in (2.55), by (2.49) we obtain immediately the value of $P^{FRA}(t; T, T + \Delta, R, N)$ and, by (2.51), also the FRA rate (fair fixed rate).

2.3.2   Adjustment Factors for FRAs

Next we want to compare the quantity $\bar{\nu}_{t,T}$ in (2.50) with the analogous quantity

$$\nu_{t,T} := E^{T+\Delta} \left\{ \frac{1}{p(T, T + \Delta)} \Big| \mathscr{F}_t \right\} = \frac{p(t, T)}{p(t, T + \Delta)}$$

(2.56)

in the single-curve case, where $p(t, T)$ are the prices of the OIS bonds and where the second equality follows from the fact that $\frac{p(\cdot, T)}{p(\cdot, T+\Delta)}$ is a martingale under the $(T + \Delta)$-forward measure. The *fair fixed rate* in the single curve case is then

$$R_t = \frac{1}{\Delta} (\nu_{t,T} - 1) = \frac{1}{\Delta} \left(\frac{p(t, T)}{p(t, T + \Delta)} - 1 \right)$$

(2.57)

Notice that, since the OIS bonds can be considered as observable (they are stripped from the OIS rates), contrary to what happens with $\bar{R}_t$, to compute R_t no interest rate model is needed. Notice also that, before the crisis, the difference between the standard zero coupon bonds and the OIS bonds was negligible. From a methodological point of view we may therefore consider the comparison of $\bar{\nu}_{t,T}$ with $\nu_{t,T}$ also as a comparison between the multi-curve (post-crisis) $\bar{\nu}_{t,T}$ and the single-curve (pre-crisis) $\nu_{t,T}$.

We shall now derive an "adjustment factor" allowing to pass from the single-curve (pre-crisis) value of $\nu_{t,T}$ to the corresponding multi-curve (post-crisis) $\bar{\nu}_{t,T}$ that bears some analogy with the "multiplicative forward basis" in Bianchetti (2010) and the multiplicative spread defined in Sect. 7.3.2 of Henrard (2014). Note that the idea of an adjustment factor in the post-crisis setup has been considered also in Piterbarg (2010), who introduces a convexity adjustment between CSA and non-CSA prices of a forward contract. We shall in fact prove the following

Proposition 2.2 *For the affine factor model classes (2.37) and (2.38), assuming in addition that $\bar{B}^3(T, T + \Delta) \in \mathscr{I}_T$ for the model class (2.37), we have*

$$\bar{\nu}_{t,T} = \nu_{t,T} \cdot Ad_t^{T,\Delta} \cdot Res_t^{T,\Delta} \tag{2.58}$$

where we shall call "adjustment factor" the second factor on the right given by

$$
\begin{aligned}
Ad_t^{T,\Delta} &:= E^Q \left\{ \frac{p(T, T + \Delta)}{\bar{p}(T, T + \Delta)} \Big| \mathscr{F}_t \right\} \\
&= e^{-\bar{A}(T,T+\Delta)} E^Q \left\{ e^{\kappa B^1(T,T+\Delta)\Psi_T^1 + \bar{B}^3(T,T+\Delta)\Psi_T^3} | \mathscr{F}_t \right\}
\end{aligned}
\tag{2.59}
$$

and the "residual factor" the last factor on the right given by

$$Res_t^{T,\Delta} = \exp\left[-\kappa \frac{(\sigma^1)^2}{2(b^1)^3} \left(1 - e^{-b^1 \Delta} \right) \left(1 - e^{-b^1(T-t)} \right)^2 \right] \tag{2.60}$$

The residual factor drops out for the case when the correlation κ between the short rate and the spread reduces to zero.

Remark 2.6 Notice that the expectation in (2.59) can be explicitly computed on the basis of Lemmas 2.1 and 2.2 as a function of the set $\theta := (a^i, b^i, \sigma^i, i = 1, 2, 3)$ of all parameters in the model, of the correlation coefficient κ and of the current values $\Psi_t^1, \Psi_t^2, \Psi_t^3$ of the factors. Recall that, for the model class (2.37), the parameters have to be such that $\bar{B}^3(T, T + \Delta) \in \mathscr{I}_T$.

Remark 2.7 Here we want to remark that, in the case when all three factors satisfy a Vasiček-type model (2.38), the adjustment factor can also be computed as an unconditional expectation. We have in fact by (2.59) and the independence of the factors

$$
\begin{aligned}
Ad_t^{T,\Delta} &= e^{-\tilde{A}(T,T+\Delta)} E^Q \left\{ \exp(\kappa B^1(T,T+\Delta)\Psi_T^1) | \mathscr{F}_t \right\} E^Q \left\{ \exp(\bar{B}^3(T,T+\Delta)\Psi_T^3) | \mathscr{F}_t \right\} \\
&= e^{-\tilde{A}(T,T+\Delta)} E^Q \left\{ \exp(\kappa B^1(T,T+\Delta)(\Psi_T^1 - \Psi_t^1)) \exp(\kappa B^1(T,T+\Delta)\Psi_t^1) | \mathscr{F}_t \right\} \\
&\quad \cdot E^Q \left\{ \exp(\bar{B}^3(T,T+\Delta)(\Psi_T^3 - \Psi_t^3)) \exp(\bar{B}^3(T,T+\Delta)\Psi_t^3) | \mathscr{F}_t \right\} \\
&= e^{-\tilde{A}(T,T+\Delta)} \exp(\kappa B^1(T,T+\Delta)\Psi_t^1 + \bar{B}^3(T,T+\Delta)\Psi_t^3) \\
&\quad \cdot E^Q \left\{ \exp(\kappa B^1(T,T+\Delta)(\Psi_T^1 - \Psi_t^1) + \bar{B}^3(T,T+\Delta)(\Psi_T^3 - \Psi_t^3)) | \mathscr{F}_t \right\}
\end{aligned}
\tag{2.61}
$$

Let now $\forall s \geq t$ and $j = 1, 3$, $Z_s^j := \Psi_s^j - \Psi_t^j$. Thus $Z_t^j = 0$ and we have

$$
dZ_s^j = d\Psi_s^j = (a^j - b^j Z_s^j - b^j \Psi_t^j)dt + \sigma^j dw_s^j
$$

so that

$$
\begin{aligned}
Z_s^j &= e^{-b^j s} \left(\frac{a^j}{b^j}(e^{b^j s} - e^{b^j t}) - \Psi_t^j (e^{b^j s} - e^{b^j t}) + \sigma^j \int_t^s e^{b^j u} dw_u^j \right) \\
&= V_s^j - \Psi_t^j (1 - e^{-b^j(s-t)})
\end{aligned}
$$

where $V_s^j := \frac{a^j}{b^j}(1 - e^{-b^j(s-t)}) + \sigma^j e^{-b^j s} \int_t^s e^{b^j u} dw_u^j$ is independent of $\mathscr{F}_t$. Consequently

$$
\begin{aligned}
Ad_t^{T,\Delta} &= e^{-\tilde{A}(T,T+\Delta)} e^{(\kappa B^1(T,T+\Delta)\Psi_t^1 + \bar{B}^3(T,T+\Delta)\Psi_t^3)} \\
&\quad \cdot e^{(-\kappa B^1(T,T+\Delta)(1 - e^{-b^1(T-t)})\Psi_t^1)} E^Q \left\{ e^{\kappa B^1(T,T+\Delta)V_T^1} \right\} \\
&\quad e^{(-\bar{B}^3(T,T+\Delta)(1 - e^{-b^3(T-t)})\Psi_t^3)} E^Q \left\{ e^{\bar{B}^3(T,T+\Delta)V_T^3} \right\}
\end{aligned}
\tag{2.62}
$$

Proof of Proposition 2.2: Recalling from (2.55) that

$$
\begin{aligned}
\bar{\nu}_{t,T} &= \frac{1}{p(t,T+\Delta)} \exp\left[-\tilde{A}(T,T+\Delta)\right] E^Q \left\{ e^{\bar{B}^3(T,T+\Delta)\Psi_T^3} | \mathscr{F}_t \right\} \\
&\quad \cdot E^Q \left\{ e^{-\int_t^T \Psi_u^2 du} | \mathscr{F}_t \right\} E^Q \left\{ e^{-\int_t^T \Psi_u^1 du + \kappa B^1(T,T+\Delta)\Psi_T^1} | \mathscr{F}_t \right\}
\end{aligned}
\tag{2.63}
$$

let us compute the last term on the right-hand side. Note that, for the model class (2.37), the first expectation on the right-hand side is finite thanks to $\bar{B}^3(T,T+\Delta) \in \mathscr{I}_T$. We have by Lemma 2.1 (with $K = -\kappa B^1(T,T+\Delta)$ and $\gamma = 1$ therein) and with $B^1(t,T)$ as given in (2.40).

$$
\begin{aligned}
&E^Q \left\{ e^{-\int_t^T \Psi_u^1 du + \kappa B^1(T,T+\Delta)\Psi_T^1} | \mathscr{F}_t \right\} \\
&= \exp\left[-B^1(t,T)\Psi_t^1 + \kappa B^1(T,T+\Delta)e^{-b^1(T-t)}\Psi_t^1 \right. \\
&\quad \left. - a^1 \int_t^T B^1(u,T)du + a^1 \kappa B^1(T,T+\Delta) \int_t^T e^{-b^1(T-u)} du \right]
\end{aligned}
$$

$$\cdot \exp\left[\frac{(\sigma^1)^2}{2} \int_t^T (B^1(u, T))^2 du + \frac{(\sigma^1)^2}{2} (\kappa B^1(T, T + \Delta))^2 \int_t^T e^{-2b^1(T-u)} du \right]$$

$$\cdot \exp\left[-\kappa B^1(T, T + \Delta)(\sigma^1)^2 \int_t^T B^1(u, T) e^{-b^1(T-u)} du \right]$$

$$= E^Q \left\{ e^{-\int_t^T \Psi_u^1 du} \mid \mathscr{F}_t \right\} E^Q \left\{ e^{\kappa B^1(T, T+\Delta)\Psi_T^1} \mid \mathscr{F}_t \right\}$$

$$\cdot \exp\left[-\kappa(\sigma^1)^2 B^1(T, T + \Delta) \int_t^T B^1(u, T) e^{-b^1(T-u)} du \right] \tag{2.64}$$

Therefore,

$$\bar{\nu}_{t,T} = \frac{1}{p(t, T + \Delta)} \exp\left[-\tilde{A}(T, T + \Delta) \right] E^Q \left\{ e^{\bar{B}^3(T, T+\Delta)\Psi_T^3} \mid \mathscr{F}_t \right\}$$

$$\cdot E^Q \left\{ e^{-\int_t^T \Psi_u^2 du} \mid \mathscr{F}_t \right\} E^Q \left\{ e^{-\int_t^T \Psi_u^1 du} \mid \mathscr{F}_t \right\} E^Q \left\{ e^{\kappa B^1(T, T+\Delta)\Psi_T^1} \mid \mathscr{F}_t \right\}$$

$$\cdot \exp\left[-\kappa(\sigma^1)^2 B^1(T, T + \Delta) \int_t^T B^1(u, T) e^{-b^1(T-u)} du \right]$$

$$= \frac{p(t, T)}{p(t, T + \Delta)} E^Q \left\{ \frac{p(T, T + \Delta)}{\bar{p}(T, T + \Delta)} \middle| \mathscr{F}_t \right\}$$

$$\cdot \exp\left[-\kappa(\sigma^1)^2 B^1(T, T + \Delta) \int_t^T B^1(u, T) e^{-b^1(T-u)} du \right] \tag{2.65}$$

by expression (2.39) for the OIS bond prices and Eq. (2.54), and we obtain

$$\bar{\nu}_{t,T} = \nu_{t,T} \cdot Ad_t^{T,\Delta} \cdot Res_t^{T,\Delta}$$

where the pre-crisis value $\nu_{t,T}$ is given by (2.56), $Ad_t^{T,\Delta}$ is defined in (2.59) and $Res_t^{T,\Delta}$ in (2.60), noticing that

$$B^1(T, T + \Delta) \int_t^T B^1(u, T) e^{-b^1(T-u)} du = \frac{1}{2(b^1)^3} \left(1 - e^{-b^1 \Delta} \right) \left(1 - e^{-b^1(T-t)} \right)^2 \tag{2.66}$$

The following immediate corollary establishes a relation between the FRA rate $\bar{R}_t$ in an actual (post-crisis) FRA and the corresponding single-curve (pre-crisis) rate R_t:

Corollary 2.1 *The following relationship holds*

$$\bar{R}_t = \left(R_t + \frac{1}{\Delta} \right) \cdot Ad_t^{T,\Delta} \cdot Res_t^{T,\Delta} - \frac{1}{\Delta} \tag{2.67}$$

Notice that the residual factor is equal to 1 for zero correlation, i.e. for $\kappa = 0$.

Remark 2.8 The adjustment factor $Ad_t^{T,\Delta}$ allows also for some intuitive interpretations. The easiest one is obtained for the case when $\kappa = 0$ (independence of

r_t and s_t): in this case we have $r_t + s_t > r_t$ implying $\bar{p}(T, T + \Delta) < p(T, T + \Delta)$ so that $Ad_t^{T, \Delta} \geq 1$. Moreover, the residual factor in (2.58) is equal to 1. As expected, from Proposition 2.2 and Corollary 2.1 it then follows that

$$\bar{\nu}_{t,T} \geq \nu_{t,T}, \quad \bar{R}_t \geq R_t \tag{2.68}$$

For some intuition in the cases when $\kappa \neq 0$ we refer to Morino and Runggaldier (2014).

2.3.2.1 Comments on Calibration to the Initial Term Structure

For what concerns calibration of the model to observed initial term structures of OIS bond prices (stripped from the OIS rates) and FRA contracts notice that this procedure can be decoupled in the following way. The coefficients $a^1, a^2, b^1, b^2, \sigma^1, \sigma^2$ can be calibrated in the usual way on the basis of the observations of OIS bond prices $p(t, T)$. To calibrate a^3, b^3, σ^3, notice that, contrary to $p(t, T)$, the fictitious bond prices $\bar{p}(t, T)$ are not observable. Considering the special case of (1.32) for $n = 1$, one can however observe the OIS FRA rates $R_t = \frac{1}{\Delta}\left(\frac{p(t,T)}{p(t,T+\Delta)} - 1\right)$, as well as the FRA rates $\bar{R}_t$. Recalling then Corollary 2.1 and the fact that $Ad_t^{T, \Delta}$ is, see Remark 2.6, a function of the type $A(\theta, \kappa, \Psi_t^1, \Psi_t^3)$ with $\theta := (a^i, b^i, \sigma^i, \ i = 1, 2, 3)$, notice that, having calibrated a^i, b^i, σ^i ($i = 1, 2$), from the observations of R_t and $\bar{R}_t$ one could thus calibrate a^3, b^3, σ^3, as well as the correlation parameter κ. Moreover, we mention that if the adjustment factor $Ad_t^{T, \Delta}$ would be directly observable, then the relationship between R_t and $\bar{R}_t$ as expressed in Corollary 2.1 would allow to easily recover κ from the residual factor $Res_t^{T, \Delta}$ defined in (2.60).

As already mentioned in Remark 2.1, a perfect fit to the initial term structure can be achieved in the model either by passing to time varying parameters, or by introducing a deterministic shift. Finally we recall that, as pointed out in Crépey et al. (2015a), calibration to clean prices is sufficient also when using the model to compute possible CVA and other valuation adjustments.

2.4 Pricing of Caps and Floors

A cap is a first basic example for a nonlinear interest rate derivative, in particular of an "optional derivative". As recalled in Sect. 1.4.6 a cap consists of a series of caplets and so we shall derive here the price for a generic caplet with strike K on the Libor rate $L(T; T, T + \Delta)$, fixed at time T for the time interval $[T, T + \Delta]$.

In Proposition 2.3 we shall derive an explicitly computable expression for the price, at $t < T$, of a caplet for a generic interval $[T, T + \Delta]$ and for the affine factor model classes (2.37) and (2.38). The price will be computed under the forward measure $Q^{T+\Delta}$, while the models are defined under the standard martingale measure Q. We shall thus make use of Lemma 2.5 that specifies the distribution under $Q^{T+\Delta}$ at a generic $t < T + \Delta$, of each of the $\Psi_t^1, \Psi_t^2, \Psi_t^3$, defined by (2.37) and (2.38).

For ease of reference, in the four points (i)–(iv) below we summarize the notation that we shall use in this subsection and that, for the points (i)–(iii), follows from Lemma 2.5. Thereby we shall assume that the processes start at a generic time $t < T$ so that all distributions and corresponding expectations below will be conditional on $\mathscr{F}_t$. To alleviate notation, we shall count time from t onwards and, in particular, set $\tau := T - t$.

(i) The factor Ψ_t^1 has, at time T and under $Q^{T+\Delta}$, a Gaussian density $N(\alpha_\tau^1, \beta_\tau^1)$ with

$$
\begin{cases}
\alpha_\tau^1 = e^{-b^1\tau}\left(\Psi_t^1 + \left(\dfrac{a^1}{b^1} - \dfrac{(\sigma^1)^2}{(b^1)^2}\right)(e^{b^1\tau} - 1) + \dfrac{(\sigma^1)^2}{2(b^1)^2}e^{-b^1(t+\tau+\Delta)}(e^{2b^1\tau} - 1)\right) \\[2mm]
\beta_\tau^1 = \dfrac{(\sigma^1)^2}{2(b^1)}e^{-2b^1\tau}(e^{2b^1\tau} - 1) > 0.
\end{cases}
$$

(2.69)

In the sequel we shall denote this density by $f_{\Psi_T^1}(\cdot) = f_1(\cdot)$.

(ii) In the model class (2.37), the factors Ψ^i ($i = 2, 3$) have, see Brigo and Mercurio (2006, Sect. 3.2.3), at time T and under $Q^{T+\Delta}$ a density $f_{\Psi_T^i}(\zeta)$ that can be expressed as equal to $c_\tau^2 g_2(c_\tau^2\zeta)$ and $c_\tau^3 g_3(c_\tau^3\zeta)$ respectively, where $g_i(\cdot)$ denotes the density of a non-central chi-square distribution with v^i degrees of freedom and non-centrality parameter λ_τ^i. The parameter values are given by (see (2.30))

$$
\begin{cases}
c_\tau^2 = \dfrac{4(b^2 + (\sigma^2)^2 B^2(\tau, t+\tau+\Delta))}{(\sigma^2)^2(1 - e^{-\tau(b^2+(\sigma^2)^2 B^2(\tau,t+\tau+\Delta))})} > 0 \\[2mm]
v^2 = \dfrac{4a^2}{(\sigma^2)^2} \\[2mm]
\lambda_\tau^2 = c_\tau^2\Psi_t^2 e^{-\tau(b^2+(\sigma^2)^2 B^2(\tau,t+\tau+\Delta))} \\[2mm]
v^3 = \dfrac{4a^3}{(\sigma^3)^2} \\[2mm]
c_\tau^3 = \dfrac{4b^3}{(\sigma^3)^2(1 - e^{-b^3\tau})} > 0 \\[2mm]
\lambda_\tau^3 = c_\tau^3\Psi_t^3 e^{-b^3\tau}
\end{cases}
$$

(2.70)

(iii) In the model class (2.38), the factors Ψ^i ($i = 2, 3$) have at time T and under $Q^{T+\Delta}$ a Gaussian density $N(\alpha_\tau^i, \beta_\tau^i)$, denoted by $f_i(\cdot)$, $i = 2, 3$, with

$$
\begin{cases}
\alpha_\tau^2 = e^{-b^2\tau}\left(\Psi_t^2 + \left(\dfrac{a^2}{b^2} - \dfrac{(\sigma^2)^2}{(b^2)^2}\right)(e^{b^2\tau} - 1) + \dfrac{(\sigma^2)^2}{2(b^2)^2}e^{-b^2(t+\tau+\Delta)}(e^{2b^2\tau} - 1)\right) \\[2mm]
\beta_\tau^2 = \dfrac{(\sigma^2)^2}{2(b^2)}e^{-2b^2\tau}(e^{2b^2\tau} - 1) \\[2mm]
\alpha_\tau^3 = e^{-b^3\tau}\left(\Psi_t^3 + \dfrac{a^3}{b^3}(e^{b^3\tau} - 1)\right) \\[2mm]
\beta_\tau^3 = \dfrac{(\sigma^3)^2}{2(b^3)}e^{-2b^3\tau}(e^{2b^3\tau} - 1)
\end{cases}
$$

(2.71)

(iv) Consider then the shorthand notations

$$\bar{A} := \bar{A}(T, T+\Delta), \ \bar{B}^1 := \bar{B}^1(T, T+\Delta), \ \bar{B}^2 := \bar{B}^2(T, T+\Delta),$$
$$\bar{B}^3 := \bar{B}^3(T, T+\Delta), \bar{K} := 1 + \Delta K$$

where $\bar{A}(\cdot)$ and $\bar{B}^j(\cdot)$, $j = 1, 2, 3$ correspond to those in (2.41) and are determined according to Lemmas 2.1 and 2.2. Finally let, for $y \in \mathbb{R}$ and $z \in \mathbb{R}$,

$$\bar{x}(y, z) = \frac{\log(\bar{K}) + \bar{A} - \bar{B}^2 y - \bar{B}^3 z}{\bar{B}^1} \qquad (2.72)$$

We can now state and prove the following

Proposition 2.3 *Assume that the correlation intensity satisfies $\kappa > -1$ and that, for the case of the factor model class (2.37), $\bar{B}^2, \bar{B}^3 \in \mathscr{I}_T^{T+\Delta}$, for $\mathscr{I}_T^{T+\Delta}$ defined in (2.16). The price of a caplet for the interval $[T, T+\Delta]$ with strike K on the Libor rate $L(T; T, T+\Delta)$ can be expressed for both model classes (2.37) and (2.38) as follows*

$$P^{Cpl}(t; T+\Delta, K) = p(t, T+\Delta) \int_{\mathbb{R}^2} e^{-\bar{A} + \bar{B}^2 y + \bar{B}^3 z}$$
$$\left[\int_{\bar{x}(y,z)}^{+\infty} (e^{\bar{B}^1 x} - e^{\bar{B}^1 \bar{x}(y,z)}) f_{\psi_T^1}(x) dx \right] f_{\psi_T^2}(y) dy f_{\psi_T^3}(z) dz$$
$$= p(t, T+\Delta)$$
$$\int_{\mathbb{R}^2} e^{-\bar{A} + \bar{B}^2 y + \bar{B}^3 z} Call(t, S_t, \bar{K}(y, z), T+\Delta) f_{\psi_T^2}(y) dy f_{\psi_T^3}(z) dz$$

$$(2.73)$$

where $Call(t, S_t, \bar{K}(y, z), T+\Delta)$ is formally the time-t price of a call option with (random) strike defined by $\bar{K}(y, z) = e^{\bar{B}^1 \bar{x}(y,z)}$, maturity $T+\Delta$ and underlying asset $S_t = e^{\bar{B}^1 \psi_t^1}$ and is given by

$$Call(t, S_t, \bar{K}, T+\Delta) = e^{(\frac{1}{2}(\bar{B}^1)^2 \beta_T^1 + \alpha_T^1 \bar{B}^1)} N \left(\frac{\alpha_T^1 + \bar{B}^1 \beta_T^1 - \bar{x}}{\sqrt{\beta_T^1}} \right) - e^{\bar{B}^1 \bar{x}(y,z)} N \left(\frac{\alpha_T^1 - \bar{x}}{\sqrt{\beta_T^1}} \right)$$

$$(2.74)$$

where $N(\cdot)$ denotes the cumulative standard Gaussian distribution function.

Proof Using the forward measure $Q^{T+\Delta}$ as pricing measure, we have to compute (see (1.42) and recall (2.41))

$$P^{Cpl}(t; T + \Delta, K) = p(t, T + \Delta)E^{T+\Delta}\left\{ \left(\frac{1}{\bar{p}(T, T + \Delta)} - \bar{K} \right)^+ \bigg| \mathscr{F}_t \right\}$$

$$= p(t, T + \Delta)E^{T+\Delta}\left\{ \left(e^{-\bar{A}+\bar{B}^1 \Psi^1_T + \bar{B}^2 \Psi^2_T + \bar{B}^3 \Psi^3_T} - \bar{K} \right)^+ \bigg| \mathscr{F}_t \right\}$$

$$= p(t, T + \Delta) \int_{\mathbb{R}^3} \left(e^{-\bar{A}+\bar{B}^1 x + \bar{B}^2 y + \bar{B}^3 z} - \bar{K} \right)^+ f_{\Psi^1_T}(x) f_{\Psi^2_T}(y) f_{\Psi^3_T}(z) dx dy dz \qquad (2.75)$$

Consider next the function

$$g(x, y, z) = e^{-\bar{A}+\bar{B}^1 x + \bar{B}^2 y + \bar{B}^3 z} \qquad (2.76)$$

For any given $y \in \mathbb{R}, z \in \mathbb{R}$, it is a monotonically increasing and continuous function of $x \in \mathbb{R}$ (notice that $\bar{B}^1 = \bar{B}^1(T, T + \Delta)$ is, according to its expression given in (2.42) together with (2.40), positive provided $\kappa > -1$) and has the property that

$$\lim_{x \to -\infty} g(x, y, z) = 0, \quad \lim_{x \to +\infty} g(x, y, z) = +\infty$$

There exists thus a unique $\bar{x}(y, z)$ such that $g(\bar{x}, y, z) = \bar{K}$ and it is given by (2.72). Furthermore,

$$x \geq \bar{x}(y, z) \Leftrightarrow g(x, y, z) \geq g(\bar{x}(y, z), y, z)$$

Consequently, (2.75) becomes

$$P^{Cpl}(t; T + \Delta, K) = p(t, T + \Delta) \int_{\mathbb{R}^2} e^{-\bar{A}+\bar{B}^2 y + \bar{B}^3 z}$$
$$\left(\int_{\bar{x}(y,z)}^{+\infty} (e^{\bar{B}^1 x} - e^{\bar{B}^1 \bar{x}(y,z)}) f_{\Psi^1_T}(x) dx \right) f_{\Psi^2_T}(y) dy f_{\Psi^3_T}(z) dz \qquad (2.77)$$

where, for the model class (2.37), the assumption $\bar{B}^2, \bar{B}^3 \in \mathscr{I}^{T+\Delta}_T$ ensures that the integrals above are finite.

For the second equality in (2.73), note that the inner integral in the equation above can be computed as

$$\int_{\bar{x}(y,z)}^{+\infty} (e^{\bar{B}^1 x} - e^{\bar{B}^1 \bar{x}(y,z)}) f_{\Psi^1_T}(x) dx = \int_{\mathbb{R}} (e^{\bar{B}^1 x} - e^{\bar{B}^1 \bar{x}(y,z)})^+ f_{\Psi^1_T}(x) dx$$
$$= E^{T+\Delta}\{(e^{\bar{B}^1 \Psi^1_T} - \bar{K}(y, z))^+ | \mathscr{F}_t\} = Call(t, S_t, \bar{K}(y, z), T + \Delta) \qquad (2.78)$$

Notice that, since we deal with caplets, although the maturity is $T + \Delta$, the underlying Ψ^1_t is evaluated at $t = T$. Taking into account that $\Psi^1_T \sim N(\alpha^1_T, \beta^1_T)$ with (α^1_T, β^1_T) given as in (2.69), $Call(t, S_t, \bar{K}(y, z), T + \Delta)$ can be explicitly computed in the following way (using the shorthand notation $\bar{x} = \bar{x}(y, z)$)

$$Call(t, S_t, \bar{K}, T + \Delta) = \int_{\bar{x}}^{+\infty} e^{\bar{B}^1 x} f_1(x) dx - \int_{\bar{x}}^{+\infty} e^{\bar{B}^1 \bar{x}} f_1(x) dx$$

$$= \int_{\bar{x}}^{+\infty} e^{\bar{B}^1 x} \frac{1}{\sqrt{2\pi \beta_T^1}} e^{-\frac{1}{2} \frac{(x - \alpha_T^1)^2}{\beta_T^1}} dx - e^{\bar{B}^1 \bar{x}} \int_{\bar{x}}^{+\infty} \frac{1}{\sqrt{2\pi \beta_T^1}} e^{-\frac{1}{2} \frac{(x - \alpha_T^1)^2}{\beta_T^1}} dx$$

$$= e^{(\frac{1}{2}(\bar{B}^1)^2 \beta_T^1 + \alpha_T^1 \bar{B}^1)} \int_{\frac{(\bar{x} - (\alpha_T^1 + \bar{B}^1 \beta_T^1))}{\sqrt{\beta_T^1}}}^{+\infty} \frac{1}{\sqrt{2\pi}} e^{-\frac{\zeta^2}{2}} d\zeta - e^{\bar{B}^1 \bar{x}} \int_{\frac{\bar{x} - \alpha_T^1}{\sqrt{\beta_T^1}}}^{+\infty} \frac{1}{\sqrt{2\pi}} e^{-\frac{\zeta^2}{2}} d\zeta$$

$$= e^{(\frac{1}{2}(\bar{B}^1)^2 \beta_T^1 + \alpha_T^1 \bar{B}^1)} N \left(\frac{\alpha_T^1 + \bar{B}^1 \beta_T^1 - \bar{x}}{\sqrt{\beta_T^1}} \right) - e^{\bar{B}^1 \bar{x}(y,z)} N \left(\frac{\alpha_T^1 - \bar{x}}{\sqrt{\beta_T^1}} \right)$$

$\square$

Note that the assumptions in the above proposition put some restrictions on the parameters of the CIR model (2.37) for the factors Ψ_t^2 and Ψ_t^3. Inserting now the density functions $f_{\Psi_T^2}(\cdot)$ and $f_{\Psi_T^3}(\cdot)$ for each model specification, we obtain

Corollary 2.2 *Under the assumptions from Proposition 2.3, the price of a caplet for the interval $[T, T + \Delta]$ with strike K on the Libor rate $L(T; T, T + \Delta)$ in the factor model (2.37) is given by*

$$P^{Cpl}(t; T + \Delta, K) = p(t, T + \Delta)$$
$$c_T^2 c_T^3 \int_{\mathbb{R}^2} e^{-\bar{A} + \bar{B}^2 y + \bar{B}^3 z} Call(t, S_t, \bar{K}(y, z), T + \Delta) g_2(c_T^2 y) dy g_3(c_T^3 z) dz \qquad (2.79)$$

where $g_2(\cdot)$ and $g_3(\cdot)$ are non-central chi-square densities as in (ii) with parameters given by (2.70), and in the model (2.38) by

$$P^{Cpl}(t; T + \Delta, K) = p(t, T + \Delta)$$
$$\int_{\mathbb{R}^2} e^{-\bar{A} + \bar{B}^2 y + \bar{B}^3 z} Call(t, S_t, \bar{K}(y, z), T + \Delta) f_2(y) dy f_3(z) dz \qquad (2.80)$$

where $f_2(\cdot)$ and $f_3(\cdot)$ are Gaussian densities as in (iii) with parameters given by (2.71).

Remark 2.9 Notice that, for both model classes (2.37) and (2.38), we have $\bar{B}^3 := \bar{B}^3(T, T + \Delta) > 0$. More precisely, for model (2.38), from (2.41) and Lemma 2.1 with $\gamma = 1$ and $K = 0$, we have $\bar{B}^3(T, T + \Delta) = -\frac{1}{b^3} \left(e^{-b^3 \Delta} - 1 \right) > 0$. For (2.37), again from (2.41) and this time from Lemma 2.2, always with $\gamma = 1$ and $K = 0$ so that $h = \sqrt{(b^3)^2 + 2(\sigma^3)^2}$, we have

$$\bar{B}^3(T, T + \Delta) = \frac{2 \left(e^{h\Delta} - 1 \right)}{h - b^3 + e^{h\Delta}(h + b^3)} = \frac{2 \left(e^{h\Delta} - 1 \right)}{2h + (b^3 + h) \left(e^{h\Delta} - 1 \right)} > 0 \quad (2.81)$$

Consequently, the function $g(x, y, z)$ in (2.76) is, for each fixed $x \in \mathbb{R}, y \in \mathbb{R}$, monotonically increasing and continuous also in $z \in \mathbb{R}$ with

$$\lim_{z \to -\infty} g(x, y, z) = 0, \quad \lim_{z \to +\infty} g(x, y, z) = +\infty$$

so that there exists a unique $\bar{z}(x, y)$ for which $g(x, y, \bar{z}) = \bar{K}$ and

$$z \geq \bar{z}(x, y) \Leftrightarrow g(x, y, z) \geq g(x, y, \bar{z}(x, y))$$

Therefore, since $g(x, y, z)$ is increasing with z, we have

$$\left(e^{-\bar{A}+\bar{B}^1 x+\bar{B}^2 y+\bar{B}^3 z} - \bar{K} \right)^+ = e^{-\bar{A}+\bar{B}^1 x+\bar{B}^2 y} \left(e^{\bar{B}^3 z} - e^{\bar{B}^3 \bar{z}} \right)$$

for $z \geq \bar{z}(x, y)$ and zero otherwise. We may then rewrite (2.77) as

$$\begin{aligned}
P^{Cpl}(t; T + \Delta, K) &= p(t, T + \Delta) \int_{\mathbb{R}^2} e^{-\bar{A}+\bar{B}^1 x+\bar{B}^2 y} \\
&\quad \left(\int_{\bar{z}(x,y)}^{+\infty} (e^{\bar{B}^3 z} - e^{\bar{B}^3 \bar{z}(x,y)}) f_{\psi_T^3}(z) dz \right) f_{\psi_T^1}(x) dx f_{\psi_T^2}(y) dy \quad (2.82) \\
&= p(t, T + \Delta) \int_{\mathbb{R}^2} e^{-\bar{A}+\bar{B}^1 x+\bar{B}^2 y} \\
&\quad Call(t, S_t, \bar{K}(x, y), T + \Delta) f_1(x) dx f_{\psi_T^2}(y) dy
\end{aligned}$$

with $\bar{K}(x, y) = e^{\bar{B}^3 \bar{z}(x,y)}$ and $S_t = e^{\bar{B}^3 \psi_t^3}$.

In the model (2.38) an alternative expression for the price $P^{Cpl}(t; T + \Delta, K)$ of the caplet can be derived, thereby exploiting the Gaussianity of the factor processes ψ^2 and ψ^3. We have in fact

Proposition 2.4 *The price of a caplet for the interval $[T, T + \Delta]$ with strike K on the Libor rate $L(T; T, T + \Delta)$ can be expressed, for the factor model (2.38) and with $\tau = T - t$, as follows*

$$P^{Cpl}(t; T + \Delta, K) = p(t, T + \Delta) \left[\exp\left(\frac{\sigma_\tau}{2} - \mu_\tau \right) N(d_\tau + \sqrt{\sigma_\tau}) - \bar{K} N(d_\tau) \right]$$
$$(2.83)$$

where $N(\cdot)$ is the cumulative standard Gaussian distribution function and

$$\begin{cases}
d_\tau &= -\dfrac{\log(\bar{K}) + \mu_\tau}{\sqrt{\sigma_\tau}} \\
\mu_\tau &= \bar{A}(T, T + \Delta) - \bar{B}^1(T, T + \Delta)\alpha_\tau^1 - \bar{B}^2(T, T + \Delta)\alpha_\tau^2 - \bar{B}^3(T, T + \Delta)\alpha_\tau^3 \\
\sigma_\tau &= (\bar{B}^1(T, T + \Delta))^2 \beta_\tau^1 + (\bar{B}^2(T, T + \Delta))^2 \beta_\tau^2 + (\bar{B}^3(T, T + \Delta))^2 \beta_\tau^3 \\
\bar{K} &= 1 + \Delta K
\end{cases}$$
$$(2.84)$$

with α_τ^1, β_τ^1 as in (2.69) and α_τ^2, β_τ^2, α_τ^3, β_τ^3 as in (2.71).

Proof We may write

$$
\begin{aligned}
P^{Cpl}(t; T+\Delta, K) &= p(t, T+\Delta)E^{T+\Delta}\left\{\left(\frac{1}{\bar{p}(T, T+\Delta)} - \bar{K}\right)^+ \Big| \mathscr{F}_t\right\} \\
&= p(t, T+\Delta)\left(E^{T+\Delta}\left\{\frac{1}{\bar{p}(T, T+\Delta)}\mathbf{1}_{\{\frac{1}{\bar{p}(T,T+\Delta)}>\bar{K}\}}\Big|\mathscr{F}_t\right\} \right. \\
&\qquad\left. -\bar{K}Q^{T+\Delta}\left\{\bar{p}(T, T+\Delta) < 1/\bar{K} \mid \mathscr{F}_t\right\}\right)
\end{aligned}
$$
(2.85)

Notice now that, since $\bar{p}(T, T+\Delta)$ is given by (2.41), by Lemmas 2.4 and 2.5, we have that it is distributed as $\exp(Y)$ with $Y \sim N(\mu_\tau, \sigma_\tau)$ where

$$
\begin{cases}
\mu_\tau = \bar{A}(T, T+\Delta) - \bar{B}^1(T, T+\Delta)\alpha_\tau^1 - \bar{B}^2(T, T+\Delta)\alpha_\tau^2 - \bar{B}^3(T, T+\Delta)\alpha_\tau^3 \\
\sigma_\tau = (\bar{B}^1(T, T+\Delta))^2\beta_\tau^1 + (\bar{B}^2(T, T+\Delta))^2\beta_\tau^2 + (\bar{B}^3(T, T+\Delta))^2\beta_\tau^3
\end{cases}
$$
(2.86)

For the second term on the right in (2.85) we then obtain

$$
\begin{aligned}
Q^{T+\Delta}\left\{\bar{p}(T, T+\Delta) < 1/\bar{K} \mid \mathscr{F}_t\right\} &= P\left[Y < \log(1/\bar{K})\right] \\
&= P\left[N(0, 1) < \frac{\log(1/\bar{K}) - \mu_\tau}{\sqrt{\sigma_\tau}}\right] = N(d_\tau).
\end{aligned}
$$
(2.87)

On the other hand, for the first term on the right in (2.85) we obtain

$$
\begin{aligned}
E^{T+\Delta}\left\{\frac{1}{\bar{p}(T, T+\Delta)}\mathbf{1}_{\{\frac{1}{\bar{p}(T,T+\Delta)}>\bar{K}\}}\Big|\mathscr{F}_t\right\} &= E^{T+\Delta}\left\{e^{-Y}\mathbf{1}_{\{e^{-Y}>\bar{K}\}} \mid \mathscr{F}_t\right\} \\
&= \int_{-\infty}^{d_\tau} \exp\left[-\sqrt{\sigma_\tau}x - \mu_\tau\right]\frac{1}{\sqrt{2\pi}}e^{-\frac{x^2}{2}}\,dx \\
&= e^{-\mu_\tau + \frac{\sigma_\tau}{2}}\int_{-\infty}^{d_\tau}\frac{1}{\sqrt{2\pi}}e^{-\frac{(x+\sqrt{\sigma_\tau})^2}{2}}\,dx = e^{-\mu_\tau + \frac{\sigma_\tau}{2}}N(d_\tau + \sqrt{\sigma_\tau})
\end{aligned}
$$
(2.88)

Inserting (2.87) and (2.88) into (2.85) we obtain the result. $\square$

2.5 Pricing of Swaptions

We first recall from Sects. 1.4.3 and 1.4.7 the most relevant aspects of a (payer) swaption as we shall use them in this section. Given a collection of payment dates $T_1 < \cdots < T_n$ with $\delta = \delta_k := T_k - T_{k-1}$, $(k = 1, \ldots, n)$ and given a fixed rate R, we first recall the price of a payer swap, initiated at $T_0 < T_1$ and evaluated at $t \leq T_0$. It is given by (see (1.27))

$$P^{Sw}(t; T_0, T_n, R) = \delta \sum_{k=1}^{n} p(t, T_k) E^{Q^k} \{L(T_{k-1}; T_{k-1}, T_k) - R | \mathscr{F}_t\}$$

$$= \delta \sum_{k=1}^{n} p(t, T_k) \left(L(t; T_{k-1}, T_k) - R\right) \qquad (2.89)$$

where, for simplicity, we have assumed the notional to be 1, i.e. $N = 1$. A swaption is then the option to enter the swap at a pre-specified initiation date T, which is thus also the maturity of the swaption and that, for simplicity of notation we assume to coincide with T_0, i.e. $T = T_0$, see Sect. 1.4.7. Its price at $t \leq T_0$ can be computed as (see the first equality in (1.43))

$$P^{Swn}(t; T_0, T_n, R) = p(t, T_0) E^{Q^{T_0}} \left\{ \left(P^{Sw}(T_0; T_n, R)\right)^+ | \mathscr{F}_t \right\} \qquad (2.90)$$

where we have used the shorthand notation $P^{Sw}(T_0; T_n, R) = P^{Sw}(T_0; T_0, T_n, R)$ and $P^{Swn}(t; T_0, T_n, R) = P^{Swn}(t; T_0, T_0, T_n, R)$. In this section we shall compute this price by first working out an explicit expression for P^{Sw} and then computing the expression on the right in (2.90).

Here we consider simultaneously the model classes (2.37) and (2.38), and as a preliminary, we shall prove Proposition 2.5, where the following shorthand notations are being used:

(a) a shorthand for various coefficients determined in Lemmas 2.1 and 2.2 for the representation of $p(T_0, T_k)$ according to (2.39) and $\bar{p}(T_{k-1}, T_k)$ according to (2.41), related to both systems (2.37) and (2.38)

$$A_k := A(T_0, T_k), \quad \bar{A}_k := \bar{A}(T_{k-1}, T_k)$$
$$B_k^j := B^j(T_0, T_k), \quad \bar{B}_k^j := \bar{B}^j(T_{k-1}, T_k), \quad j = 1, 2, 3$$

(b) a series of coefficients $\bar{\alpha}^j(t, T_{k-1})$, $\bar{\beta}^j(t, T_{k-1})$ appearing in the representations

$$E^{Q^k} \{e^{\bar{B}_k^j \Psi_{T_{k-1}}^j} | \mathscr{F}_t\} = e^{\bar{\alpha}^j(t, T_{k-1}) - \bar{\beta}^j(t, T_{k-1}) \Psi_t^j}, \quad j = 1, 2, 3$$

Furthermore

$$D^k := e^{-\bar{A}_k} \exp[\bar{\alpha}^1(T_0, T_{k-1}) + \bar{\alpha}^2(T_0, T_{k-1}) + \bar{\alpha}^3(T_0, T_{k-1})]$$

We have in both models (2.37) and (2.38)

$$\bar{\beta}^1(t, T_{k-1}) = -\bar{B}_k^1 e^{-b^1(T_{k-1}-t)}$$

$$\bar{\alpha}^1(t, T_{k-1}) = -(a^1 - (\sigma^1)^2 B^1(t, T_k)) \int_t^{T_{k-1}} \bar{\beta}^1(u, T_{k-1}) du$$

$$+ \frac{(\sigma^1)^2}{2} \int_t^{T_{k-1}} (\bar{\beta}^1(u, T_{k-1}))^2 du$$

For (2.37) we have, furthermore,

$$\bar{\beta}_t^2(t, T_{k-1}) - \bar{\beta}^2(t, T_{k-1})[b^2 + (\sigma^2)^2 B^2(t, T_k)] - \tfrac{1}{2}(\sigma^2)^2(\bar{\beta}^2(t, T_{k-1}))^2 = 0$$
$$\bar{\beta}^2(T_{k-1}, T_{k-1}) = -B_k^2$$
$$\bar{\alpha}^2(t, T_{k-1}) = -a^2 \int_t^{T_{k-1}} \bar{\beta}^2(u, T_{k-1}) du$$
$$\bar{\beta}^3(t, T_{k-1}) = \frac{-2b^3 \bar{B}_k^3}{-B_k^3(\sigma^3)^2(e^{b^3(T_{k-1}-t)}-1)+2b^3 e^{b^3(T_{k-1}-t)}}$$
$$\bar{\alpha}^3(t, T_{k-1}) = -a^3 \int_t^{T_{k-1}} \bar{\beta}^3(u, T_{k-1}) du$$

and, for (2.38),

$$\bar{\beta}^2(t, T_{k-1}) = -\bar{B}_k^2 e^{-b^2(T_{k-1}-t)}$$

$$\bar{\alpha}^2(t, T_{k-1}) = -(a^2 - (\sigma^2)^2 B^2(t, T_k)) \int_t^{T_{k-1}} \bar{\beta}^2(u, T_{k-1}) du$$
$$+ \frac{(\sigma^2)^2}{2} \int_t^{T_{k-1}} (\bar{\beta}^2(u, T_{k-1}))^2 du$$

$$\bar{\beta}^3(t, T_{k-1}) = -\bar{B}_k^3 e^{-b^3(T_{k-1}-t)}$$

$$\bar{\alpha}^3(t, T_{k-1}) = -a^3 \int_t^{T_{k-1}} \bar{\beta}^3(u, T_{k-1}) du + \frac{(\sigma^3)^2}{2} \int_t^{T_{k-1}} (\bar{\beta}^3(u, T_{k-1}))^2 du$$

(c) finally

$$\tilde{B}_k^j := \bar{\beta}^j(T_0, T_{k-1}) + B_k^j, \quad j = 1, 2, \quad \tilde{B}_k^3 := \bar{\beta}^3(T_0, T_{k-1})$$

We have now

Proposition 2.5 *Consider both model classes (2.37) and (2.38) and, for the class (2.37), assume that $\bar{B}_k^2, \bar{B}_k^3 \in \mathscr{I}_{T_{k-1}}^{T_k}$, for $\mathscr{I}_{T_{k-1}}^{T_k}$ defined in (2.16). The price of the swap according to (2.89) is given at the inception time T_0 by*

$$P^{Sw}(T_0; T_n, R) = \sum_{k=1}^{n} \left[D^k e^{A_k} e^{-\Psi_{T_0}^1 \tilde{B}_k^1 - \Psi_{T_0}^2 \tilde{B}_k^2 - \Psi_{T_0}^3 \tilde{B}_k^3} - (R\delta + 1) e^{A_k} e^{-\Psi_{T_0}^1 B_k^1 - \Psi_{T_0}^2 B_k^2} \right] \quad (2.91)$$

Proof Considering the first equality in (2.89), the crucial quantity to compute is

$$E^{Q^k}\{\delta L(T_{k-1}; T_{k-1}, T_k)|\mathscr{F}_{T_0}\} = E^{Q^k}\left\{ \frac{1}{p(T_{k-1}, T_k)} \Big| \mathscr{F}_{T_0} \right\} - 1$$
$$= \exp(-\bar{A}_k) \prod_{j=1}^{3} E^{Q^k}\{\exp(\bar{B}_k^j \Psi_{T_{k-1}}^j)|\mathscr{F}_{T_0}\} - 1$$
$$= D^k \exp[-\bar{\beta}^1(T_0, T_{k-1})\Psi_{T_0}^1 - \bar{\beta}^2(T_0, T_{k-1})\Psi_{T_0}^2 - \bar{\beta}^3(T_0, T_{k-1})\Psi_{T_0}^3] - 1 \quad (2.92)$$

where the second equality follows by (2.41). For the third equality we first apply Lemma 2.3 to express the factors $\Psi_{T_{k-1}}^j$, $j = 1, 2, 3$, under the forward measure Q^k and then we apply Lemmas 2.1 and 2.2 with $K = -\bar{B}_k^j$ (note that $\bar{B}_k^2, \bar{B}_k^3 \in \mathscr{I}_{T_{k-1}}^{T_k}$ by assumption), $j = 1, 2, 3$, to obtain the coefficients $\bar{\alpha}^j(t, T_{k-1})$, $\bar{\beta}^j(t, T_{k-1})$ in the affine representation of the conditional expectation. The shorthand notations in (a), (b), (c) above have been used.

Combining this with the representation of the bond price $p(T_0, T_k)$ given in (2.39), we obtain the result (2.91). □

Let us now derive the expressions for the swaption price. Denote

$$
\begin{cases}
g(x, y, z) = \sum_{k=1}^n D^k e^{A_k} \exp[-\tilde{B}_k^1 x - \tilde{B}_k^2 y - \tilde{B}_k^3 z] \\
h(x, y) = (R\delta + 1) \sum_{k=1}^n e^{A_k} \exp[-B_k^1 x - B_k^2 y]
\end{cases}
\tag{2.93}
$$

Below, when mentioning a density, we shall mean the density conditional on $\mathscr{F}_t$.

Proposition 2.6 *Let the assumptions from Proposition 2.5 be satisfied and for the model class (2.37) we assume in addition that $b^3 > \frac{\sigma^3}{2}$ and $-\tilde{B}_k^2, -\tilde{B}_k^3 \in \mathscr{I}_{T_0}^{T_0}$, for $\mathscr{I}_{T_0}^{T_0}$ defined in (2.16). The value of $P^{Swn}(t; T_0, T_n, R)$ as defined in (2.90) is given for both model classes (2.37) and (2.38) by*

$$
P^{Swn}(t; T_0, T_n, R) = p(t, T_0) \sum_{k=1}^n e^{A_k} D^k \int_{\mathbb{R}^2} \exp\left[-\tilde{B}_k^1 x - \tilde{B}_k^2 y\right]
$$
$$
\cdot \left(\int_{\bar{z}(x,y)}^{+\infty} \left[e^{-\tilde{B}_k^3 z} - e^{-\tilde{B}_k^3 \bar{z}(x,y)} \right] f_{\psi_{T_0}^3}(z) dz \right) f_{\psi_{T_0}^2}(y) f_{\psi_{T_0}^1}(x) dy dx
\tag{2.94}
$$

where $\bar{z}(x, y)$ is the unique solution of the equation $g(x, y, z) = h(x, y)$ for $x, y, z \in \mathbb{R}^3$.

Proof Notice that from (2.90) and (2.91) we have

$$
P^{Swn}(t; T_0, T_n, R) = p(t, T_0)
$$
$$
\int_{\mathbb{R}^3} \left[\sum_{k=1}^n D^k e^{A_k} \exp(-\tilde{B}_k^1 x - \tilde{B}_k^2 y - \tilde{B}_k^3 z) \right.
$$
$$
\left. - \sum_{k=1}^n (R\delta + 1) e^{A_k} \exp(-B_k^1 x - B_k^2 y) \right]^+ f_{\psi_{T_0}^1}(x) f_{\psi_{T_0}^2}(y) f_{\psi_{T_0}^3}(z) dx dy dz
\tag{2.95}
$$

where we have used the fact that the factor processes $\Psi^j, j = 1, 2, 3$, are independent.

Notice next that $\tilde{B}_k^3 = \bar{\beta}^3(T_0, T_{k-1}) < 0$, for each $k = 1, \ldots, n$, provided in the model class (2.37) we have $b^3 > \frac{\sigma^3}{2}$. In fact, from Remark 2.9, setting $\delta = T_k - T_{k-1}$, we first have that $\bar{B}_k^3 > 0$ for both model classes. Now, for the model class (2.38) we simply have (see point (b) above)

$$
\bar{\beta}^3(t, T_{k-1}) = -\bar{B}_k^3 e^{-b^3(T_{k-1}-t)} < 0
$$

On the other hand, for the model class (2.37) recall first from Remark 2.9 that, always for $\delta = T_k - T_{k-1}$ and $h = \sqrt{(b^3)^2 + 2(\sigma^3)^2}$,

$$
\bar{B}_k^3 = \frac{2\left(e^{h\delta} - 1\right)}{2h + (b^3 + h)\left(e^{h\delta} - 1\right)} > 0
\tag{2.96}
$$

Furthermore, from point (b) above we have that, being $\bar{B}_k^3 > 0$, the property $\bar{\beta}^3(t, T_{k-1}) < 0$ is equivalent to requiring that $(\sigma^3)^2 \bar{B}_k^3 - 2b^3 < 0$ which, given (2.96), becomes equivalent to

$$2(\sigma^3)^2 \left(e^{h\delta} - 1\right) < 4b^3 h + 2\left[(b^3)^2 + b^3 h\right]\left(e^{h\delta} - 1\right) \tag{2.97}$$

Now, if according to the assumption we have $b^3 > \frac{\sigma^3}{2}$, it follows that

$$(b^3)^2 + b^3 h = (b^3)^2 + b^3 \sqrt{(b^3)^2 + 2(\sigma^3)^2} > (\sigma^3)^2 \left[\frac{1}{4} + \frac{1}{2}\sqrt{\frac{1}{4} + 2}\right] = (\sigma^3)^2$$

implying that

$$2(\sigma^3)^2 \left(e^{h\delta} - 1\right) < 2\left[(b^3)^2 + b^3 h\right]\left(e^{h\delta} - 1\right)$$

and thus, a fortiori, (2.97).

It follows that, for each $x, y \in \mathbb{R}$, the function $g(x, y, z)$ in (2.93) is monotonically increasing and continuous in $z \in \mathbb{R}$ with

$$\lim_{z \to -\infty} g(x, y, z) = 0 \quad, \quad \lim_{z \to +\infty} g(x, y, z) = +\infty$$

Analogously to Proposition 2.3 and Remark 2.9 there exists thus a unique $\bar{z}(x, y)$ for which $g(x, y, \bar{z}) = h(x, y)$ and

$$z \geq \bar{z}(x, y) \Leftrightarrow g(x, y, z) \geq g(x, y, \bar{z}(x, y))$$

Here z corresponds to the factor Ψ_t^3 and we have

$$\left[\sum_{k=1}^{n} D^k e^{A_k} \exp(-\tilde{B}_k^1 x - \tilde{B}_k^2 y - \tilde{B}_k^3 z) - \sum_{k=1}^{n} (R\delta + 1) e^{A_k} \exp(-B_k^1 x - B_k^2 y)\right]^+$$
$$= \sum_{k=1}^{n} D^k e^{A_k} \exp[-\tilde{B}_k^1 x - \tilde{B}_k^2 y]\left(e^{-\tilde{B}_k^3 z} - e^{-\tilde{B}_k^3 \bar{z}(x, y)}\right)$$

for $z \geq \bar{z}(x, y)$. We may then continue (2.95) as

$$P^{Swn}(t; T_0, T_n, R) = p(t, T_0) \int_{\mathbb{R}^2} \left[\sum_{k=1}^{n} D^k e^{A_k} \exp\left[-\tilde{B}_k^1 x - \tilde{B}_k^2 y\right]\right.$$
$$\left.\left(\int_{\bar{z}(x,y)}^{\infty} \left[e^{-\tilde{B}_k^3 z} - e^{-\tilde{B}_k^3 \bar{z}(x,y)}\right] f_{\Psi_{T_0}^3}(z) dz\right)\right] f_{\Psi_{T_0}^1}(x) f_{\Psi_{T_0}^2}(y) dy dx \tag{2.98}$$

which is exactly (2.94). Note that all the integrals above are finite; in the model class (2.37) this is guaranteed by the assumptions. $\qquad \square$

Remark 2.10 Proposition 2.6 is based on solving the equation $g(x, y, z) = h(x, y)$ with respect to z. For caps we had in Sect. 2.4 a related equation, namely $g(x, y, z) = \bar{K}$, that was solved with respect to x (see (2.72)). There, in line with Remark 2.9, we could equivalently have chosen a $\bar{z}(x, y)$ to solve the same equation.

Here we cannot choose by analogy to the caps an $\bar{x}(y, z)$ since we set $g(x, y, z)$ not equal to a constant, but to a function $h(x, y)$ of x, y. The reason why here we have to consider the function $h(x, y)$ comes from the fact that the fixed rate R is here multiplied by $\sum_{k=1}^{n} \delta p(t, T_k)$ and the $p(t, T_k)$ depend on $\Psi_{T_0}^1$ and $\Psi_{T_0}^2$, see (2.89).

Let us now give more explicit representations of the swaption price for each model class. We begin with the model class (2.37). Similarly as for caplets, note that the assumptions from Propositions 2.5 and 2.6 put some restrictions on the parameters of the model (2.37) for what concerns the CIR equations for Ψ_t^2 and Ψ_t^3.

Corollary 2.3 *Under the assumptions from Propositions 2.5 and 2.6, the swaption price $P^{Swn}(t; T_0, T_n, R)$ for the model class (2.37) is given by*

$$P^{Swn}(t; T_0, T_n, R) = p(t, T_0)c_{\tau_0}^2 c_{\tau_0}^3 \sum_{k=1}^{n} e^{A_k} D^k \int_{\mathbb{R}^2} \exp\left[-\tilde{B}_k^1 x - \tilde{B}_k^2 y\right]$$
$$\cdot \left(\int_{\bar{z}(x,y)}^{+\infty} \left[e^{-\tilde{B}_k^3 z} - e^{-\tilde{B}_k^3 \bar{z}(x,y)}\right] g_3(c_{\tau_0}^3 z)dz\right) g_2(c_{\tau_0}^2 y)f_1(x)dydx \tag{2.99}$$

where $\bar{z}(x, y)$ is the unique solution of the equation $g(x, y, z) = h(x, y)$ for $x, y, z \in \mathbb{R}$ and $f_1(\cdot)$, $g_2(\cdot)$ and $g_3(\cdot)$ are as in points (i) and (ii) of Sect. 2.4, with

$$\tau_0 := T_0 - t \ ; \ c_{\tau_0}^2 = \frac{4(b^2 + (\sigma^2)^2 B^2(\tau_0, t+\tau_0))}{(\sigma^2)^2 \left(1 - e^{\tau_0(b^2+(\sigma^2)^2 B^2(\tau_0, t+\tau_0))}\right)} \ ; \tag{2.100}$$
$$\lambda_{\tau_0}^2 = c_{\tau_0}^2 \Psi_t^2 e^{-\tau_0(b^2+(\sigma^2)^2 B^2(\tau_0, t+\tau_0))}$$

and $c_{\tau_0}^3$ and $\lambda_{\tau_0}^3$ correspond to the above formulas for $c_{\tau_0}^2$ and $\lambda_{\tau_0}^2$ by changing all superscripts to 3 and setting the factor $B^2(\tau_0, t + \tau_0) = 0$.

By assuming now the Vasiček model (2.38) for all three factors we do not have to impose any restrictions on the parameters and also we can obtain a more explicit formula for the swaption price due to the Gaussianity of the factors. We have

Corollary 2.4 *For the model class (2.38), the swaption price $P^{Swn}(t; T_0, T_n, R)$ is given by*

$$P^{Swn}(t; T_0, T_n, R) = p(t, T_0) \sum_{k=1}^{n} D^k e^{A_k} \int_{\mathbb{R}^2} e^{-x\tilde{B}_k^1 - y\tilde{B}_k^2}$$
$$\left[e^{\frac{1}{2}(\tilde{B}_k^3)^2 \beta^3 - \alpha^3 \tilde{B}_k^3} N\left(\frac{\alpha^3 - \tilde{B}_k^3 \beta^3 - \bar{z}(x,y)}{\sqrt{\beta^3}}\right) - e^{-\tilde{B}_k^3 \bar{z}(x,y)} N\left(\frac{\alpha^3 - \bar{z}(x,y)}{\sqrt{\beta^3}}\right)\right] f_2(y)f_1(x)dydx \tag{2.101}$$

where $N(\cdot)$ is the cumulative standard Gaussian distribution function, $\alpha^1 = \alpha_{\tau_0}^1$, $\beta^1 = \beta_{\tau_0}^1$ with $\alpha_{\tau_0}^1, \beta_{\tau_0}^1$ as given in Eq. (2.69) of (i) in Sect. 2.4. Furthermore, for $j = 2, 3$, $f_j(\cdot)$, are Gaussian densities with parameters $\alpha^j = \alpha_{\tau_0}^j$, $\beta^j = \beta_{\tau_0}^j$ with $\alpha_{\tau_0}^j, \beta_{\tau_0}^j$ as given in (2.71).

Proof Since the factors $\Psi_{T_0}^j$, $j = 1, 2, 3$, have Gaussian distribution, the swaption price given in (2.94) can be further computed as

$$
P^{Swn}(t; T_0, T_n, R) = p(t, T_0) \sum_{k=1}^n e^{A_k} D^k \int_{\mathbb{R}^2} \exp\left[-\tilde{B}_k^1 x - \tilde{B}_k^2 y\right]
$$
$$
\cdot \left(\int_{\bar{z}(x,y)}^{+\infty} \left[e^{-\tilde{B}_k^3 z} - e^{-\tilde{B}_k^3 \bar{z}(x,y)}\right] f_3(z) dz\right) f_2(y) f_1(x) dy dx
\tag{2.102}
$$

where $f_j(\cdot) \sim N(\alpha^j, \beta^j)$ is a Gaussian density with $\alpha^j = \alpha_{T_0}^j$, $\beta^j = \beta_{T_0}^j$ where $\alpha_{T_0}^j$, $\beta_{T_0}^j$ are given as in (2.69) and (2.71).

Due to the fact that $f_3(\cdot)$ is the density of a Gaussian distribution $N(\alpha^3, \beta^3)$, the integral in the second line can be computed explicitly and we have straightforwardly

$$
\int_{\bar{z}(x,y)}^{+\infty} e^{-\tilde{B}_k^3 z} f_3(z) dz = e^{\frac{1}{2}(-\tilde{B}_k^3)^2 \beta^3 - \alpha^3 \tilde{B}_k^3} \int_{\frac{\bar{z}(x,y)-(\alpha^3-\tilde{B}_k^3\beta^3)}{\sqrt{\beta^3}}}^{+\infty} \frac{1}{\sqrt{2\pi}} e^{-\frac{\zeta^2}{2}} d\zeta
$$

and

$$
\int_{\bar{z}(x,y)}^{+\infty} e^{-\tilde{B}_k^3 \bar{z}(x,y)} f_3(z) dz = e^{-\tilde{B}_k^3 \bar{z}(x,y)} \int_{\frac{\bar{z}(x,y)-\alpha^3}{\sqrt{\beta^3}}}^{+\infty} \frac{1}{\sqrt{2\pi}} e^{-\frac{\zeta^2}{2}} d\zeta
$$

thus leading to the conclusion. $\qquad\square$

Remark 2.11 It is interesting to note that the price of a swaption can be obtained as linear combination of expressions that may be interpreted, at least formally, as caplet prices. In fact, for both model classes (2.37) and (2.38), starting from formula (2.94) for the price of the swaption, we can express the inner integrals as

$$
\int_{\bar{z}(x,y)}^{+\infty} e^{-\tilde{B}_k^3 z} f_3(z) dz - \int_{\bar{z}(x,y)}^{+\infty} e^{-\tilde{B}_k^3 \bar{z}(x,y)} f_3(z) dz
$$
$$
= \int_{\mathbb{R}} \left(e^{-\tilde{B}_k^3 z} - e^{-\tilde{B}_k^3 \bar{z}(x,y)}\right)^+ f_{\Psi_{T_0}^3}(z) dz = E^{Q_{T_0}}\left\{(e^{-\tilde{B}_3^k \Psi_{T_0}^3} - \bar{K}^k(x,y))^+ \mid \mathscr{F}_t\right\}
$$
$$
= Call(t, S_t^k, \bar{K}^k(x,y), T_0)
\tag{2.103}
$$

where, see also Proposition 2.3, $Call(t, S_t^k, \bar{K}^k(x, y), T_0)$ is formally the price at t of a call option with maturity T_0, underlying $S_t^k = e^{-\tilde{B}_3^k \Psi_t^3}$ and (random) strike $\bar{K}^k(x, y) = e^{-\tilde{B}_3^k \bar{z}(x,y)}$. The second equality follows from the monotonicity of the exponential function and the fact that $\tilde{B}_k^3 < 0$, as shown after Eq. (2.95).

Formula (2.94) then writes as follows

$$
P^{Swn}(t; T_0, T_n, R) = p(t, T_0)\left[\sum_{k=1}^n D^k e^{A_k} \int_{\mathbb{R}^2} \exp(-\tilde{B}_k^1 x - \tilde{B}_k^2 y)\right.
$$
$$
\left. Call(t, S_t^k, \bar{K}^k(x, y), T_0) f_{\Psi_{T_0}^1}(x) f_{\Psi_{T_0}^2}(y) dx dy\right]
\tag{2.104}
$$

On the other hand, recall from Remark 2.9, Eq. (2.82), that the price of a caplet can be written as

$$P^{Cpl}(t; T + \Delta, K) = p(t, T + \Delta) \int_{\mathbb{R}^2} e^{-\bar{A} + \bar{B}^1 x + \bar{B}^2 y} \, Call(t, S_t, \bar{K}(x, y), T + \Delta)$$

$$f_{\psi_T^1}(x) f_{\psi_T^2}(y) dx dy$$

$$:= Cpl\left(\bar{A}, \bar{B}^1, \bar{B}^2, \bar{B}^3, \bar{z}(x, y), t, T + \Delta\right) \qquad (2.105)$$

With the arguments in the function $Call(\cdot)$ given here by $S_t = e^{\bar{B}^3 \psi_t^3}$, $\bar{K}(x, y) = e^{\bar{B}_3^k \bar{z}(x,y)}$, formula (2.104) becomes the following expression as a linear combination

$$P^{Swn}(t; T_0, T_n, R) = \sum_{k=1}^{n} D^k Cpl\left(-A_k, -\tilde{B}_k^1, -\tilde{B}_k^2, -\tilde{B}_k^3, \bar{z}(x, y), t, T_0\right) \qquad (2.106)$$

We conclude this subsection with an alternative swaption pricing formula that exploits the Gaussianity of the factor processes and corresponds to Proposition 2.4 for the caplets.

Proposition 2.7 *For the model class (2.38) the value $P^{Swn}(t; T_0, T_n, R)$ can also be expressed as*

$$P^{Swn}(t; T_0, T_n, R)$$

$$= p(t, T_0) \sum_{k=1}^{n} \left[e^{A_k} \int_{\mathbb{R}^2} \left(D^k e^{-\tilde{B}_k^1 x - \tilde{B}_k^2 y} e^{\frac{1}{2}(\tilde{B}_k^3)^2 \beta^3 - \alpha^3 \tilde{B}_k^3} N\left(\frac{\alpha^3 - \tilde{B}_k^3 \beta^3 - \bar{z}(x, y)}{\sqrt{\beta^3}} \right) \right. \right.$$

$$\left. \left. - (R\delta + 1) \exp(-B_k^1 x - B_k^2 y) N\left(\frac{\alpha^3 - \bar{z}(x, y)}{\sqrt{\beta^3}} \right) \right) f_1(x) f_2(y) dx dy \right] \qquad (2.107)$$

Proof Starting from (2.102), and recalling the definition of $\bar{z}(x, y)$ in Proposition 2.6 as well as that of $h(x, y)$ in (2.93), we may alternatively write

$$P^{Swn}(t; T_0, T_n, R)$$

$$= p(t, T_0) \sum_{k=1}^{n} \left[D^k e^{A_k} \int_{\mathbb{R}^2} \exp(-\tilde{B}_k^1 x - \tilde{B}_k^2 y) \right.$$

$$\left(\int_{\bar{z}(x,y)}^{+\infty} e^{-\tilde{B}_k^3 z} f_3(z) dz \right) f_1(x) f_2(y) dx dy$$

$$\left. - (R\delta + 1) e^{A_k} \int_{\mathbb{R}^2} \exp(-B_k^1 x - B_k^2 y) \left(\int_{\bar{z}(x,y)}^{+\infty} f_3(z) dz \right) f_1(x) f_2(y) dx dy \right] \qquad (2.108)$$

and from here one obtains the conclusion by computing the integrals with respect to z on the right as was done in the proof of Corollary 2.4. $\qquad \square$

2.6 Relationship with Models from the Literature

Here we briefly discuss some of the main short-rate models, which in the context of multi-curves have appeared in the literature and show how they can be seen as particular cases of our affine factor model of Sect. 2.1, including the Gaussian, exponentially quadratic model of Sect. 2.2.

2.6.1 The Models of Kenyon (2010) and Kijima et al. (2009)

Kenyon (2010) considers for the short rate, called there $r_D(t)$, a G1++ type model (these models are recalled in Remark 2.1) of the form

$$r_D(t) = \Psi(t) + \phi_D(t)$$

where $\Psi(t)$ satisfies a Vašiček equation of the kind of the first equation in (2.3) (equivalently (2.37)) with $a^1 = 0$. The author then considers a "fixing curve" $r_{f^\Delta}(t)$ that corresponds to $r_t + s_t$ in our case and for this curve he also assumes a G1++ model. He points out how the short-rate modeling has a strong connection to intensity-based credit risk modeling.

Kijima et al. (2009) consider three short rates associated to three "discounting curves" characterized by the pedixes D (for discounting cash flows), L (relating to Libor), G (relating to government bonds). They model separately the short rate $r_D(t)$ and the spreads $h_L(t) := r_L(t) - r_D(t)$, $h_G(t) := r_G(t) - r_D(t)$. Their process $r_D(t)$ may be considered to correspond to our r_t, $r_L(t)$ to our s_t and $r_L(t) - r_G(t) = h_L(t) - h_G(t)$ to our ρ_t (see (2.2)). They consider two model setups: (i) a quadratic Gaussian model for $r_D(t)$ (corresponds to what we call Gaussian exponentially quadratic, see Sect. 2.2) and a Vašiček model for $h_L(t)$ and $h_G(t)$. To derive formulas in closed form in this first setup, they have to assume short rate and spreads to be mutually independent. To allow also for dependence they then consider the second setup (ii): short rate and spreads all satisfy a Vašiček model with correlated noises. Their model can thus be fully included in our setup.

2.6.2 The Model of Filipović and Trolle (2013)

The short rate and the spread dynamics of the paper Filipović and Trolle (2013) are described in their Sect. 3.2.

We start by showing that their short-rate model (26) can be obtained as particular case of the affine factor model for the short rate presented in Sect. 2.1.3. For this purpose we start from a two-factor model where each of the factors satisfies a pure mean-reverting Gaussian model as in (2.38) namely

$$d\Psi_t^i = (a^i - b^i \Psi_t^i)dt + \sigma^i \, dw_t^i, \quad i = 1, 2 \tag{2.109}$$

with independent Wiener processes w_t^1, w_t^2. For $\gamma^1, \gamma^2 \in \mathbb{R}^+$ put

$$r_t = \gamma^1 \Psi_t^1 + \gamma^2 \Psi_t^2 \tag{2.110}$$

Given the two Wiener processes w_t^1, w_t^2, define $\bar{w}_t^1$ as

$$d\bar{w}_t^1 = \frac{\gamma^1 \sigma^1 dw_t^1 + \gamma^2 \sigma^2 dw_t^2}{\sqrt{(\gamma^1 \sigma^1)^2 + (\gamma^2 \sigma^2)^2}} \tag{2.111}$$

From (2.110) we can then write

$$\begin{aligned}
dr_t &= \left[(\gamma^1 a^1 + \gamma^2 a^2) - b^1(\gamma^1 \Psi_t^1 + \gamma^2 \Psi_t^2) + (b^1 - b^2)\gamma^2 \Psi_t^2 \right] dt \\
&\quad + \sqrt{(\gamma^1 \sigma^1)^2 + (\gamma^2 \sigma^2)^2} d\bar{w}_t^1 \\
&= b^1 \left[\frac{\gamma^1 a^1 + \gamma^2 a^2}{b^1} + \frac{b^1 - b^2}{b^1} \gamma^2 \Psi_t^2 - r_t \right] dt + \sqrt{(\gamma^1 \sigma^1)^2 + (\gamma^2 \sigma^2)^2} d\bar{w}_t^1
\end{aligned} \tag{2.112}$$

Define next

$$\gamma_t := \frac{\gamma^1 a^1 + \gamma^2 a^2}{b^1} + \frac{b^1 - b^2}{b^1} \gamma^2 \Psi_t^2 \tag{2.113}$$

so that

$$\begin{aligned}
d\gamma_t &= \frac{b^1 - b^2}{b^1} \gamma^2 \left[(a^2 - b^2 \Psi_t^2)dt + \sigma^2 \, dw_t^2 \right] \\
&= \frac{b^1 - b^2}{b^1} \gamma^2 \left[\left(a^2 + \frac{b^1 b^2}{\gamma^2 (b^1 - b^2)} \frac{\gamma^1 a^1 + \gamma^2 a^2}{b^1} \right) - \frac{b^1 b^2}{\gamma^2 (b^1 - b^2)} \gamma_t \right] dt \\
&\quad + \frac{b^1 - b^2}{b^1} \gamma^2 \sigma^2 \, dw_t^2 \\
&:= \kappa_\gamma [\theta_\gamma - \gamma_t] \, dt + \sigma_\gamma dw_t^2
\end{aligned} \tag{2.114}$$

with $\kappa_\gamma = b^2$, $\theta_\gamma = \frac{a^2 \gamma^2 (b^1 - b^2)}{b^1 b^2} + \frac{\gamma^1 a^1 + \gamma^2 a^2}{b^1}$, $\sigma_\gamma = \frac{b^1 - b^2}{b^1} \gamma^2 \sigma^2$ while from (2.112) and (2.113) we get

$$dr_t = \kappa_r [\gamma_t - r_t] \, dt + \sigma_r d\bar{w}_t^1 \tag{2.115}$$

with $\kappa_r = b^1$, $\sigma_r = \sqrt{(\gamma^1 \sigma^1)^2 + (\gamma^2 \sigma^2)^2}$.

Relations (2.115) and (2.114) come close, but are not yet equal to (26) in Filipović and Trolle (2013), in particular $\bar{w}_t^1$ and w_t^2 are not independent.

On the other hand we have independence between $\bar{w}_t^1$ and the following process $\bar{w}_t^2$ defined as

$$\bar{w}_t^2 := \frac{\sqrt{(\gamma^1\sigma^1)^2 + (\gamma^2\sigma^2)^2}}{\sigma^2\gamma^2} \left(w_t^1 - \frac{\sigma^1\gamma^1}{\sqrt{(\gamma^1\sigma^1)^2 + (\gamma^2\sigma^2)^2}} \bar{w}_t^1 \right) \qquad (2.116)$$

In fact, both are Gaussian random processes and

$$E\left\{ \bar{w}_t^1\, \bar{w}_t^2 \right\} = \frac{\sigma^1\gamma^1}{\sigma^2\gamma^2}\, t - \frac{\sigma^1\gamma^1}{\sigma^2\gamma^2}\, t = 0$$

From (2.116) we may now write

$$w_t^1 = \frac{\sigma^2\gamma^2}{\sqrt{(\gamma^1\sigma^1)^2 + (\gamma^2\sigma^2)^2}}\, \bar{w}_t^2 + \frac{\sigma^1\gamma^1}{\sqrt{(\gamma^1\sigma^1)^2 + (\gamma^2\sigma^2)^2}}\, \bar{w}_t^1 \qquad (2.117)$$

so that

$$\frac{\sigma^1\gamma^1}{\sigma^2\gamma^2}\, dw_t^1 = \frac{\sigma^1\gamma^1}{\sqrt{(\gamma^1\sigma^1)^2 + (\gamma^2\sigma^2)^2}}\, d\bar{w}_t^2 + \frac{(\sigma^1\gamma^1)^2}{\sigma^2\gamma^2\sqrt{(\gamma^1\sigma^1)^2 + (\gamma^2\sigma^2)^2}}\, d\bar{w}_t^1 \quad (2.118)$$

On the other hand, from (2.111) we have

$$dw_t^2 = \frac{1}{\sigma^2\gamma^2} \left(\sqrt{(\gamma^1\sigma^1)^2 + (\gamma^2\sigma^2)^2}\, d\bar{w}_t^1 - \sigma^1\gamma^1\, dw_t^1 \right)$$

which, combined with (2.118), leads to

$$dw_t^2 = \left[\frac{\sqrt{(\gamma^1\sigma^1)^2 + (\gamma^2\sigma^2)^2}}{\sigma^2\gamma^2} - \frac{(\sigma^1\gamma^1)^2}{\sigma^2\gamma^2\sqrt{(\gamma^1\sigma^1)^2 + (\gamma^2\sigma^2)^2}} \right] d\bar{w}_t^1 - \frac{\sigma^1\gamma^1}{\sqrt{(\gamma^1\sigma^1)^2 + (\gamma^2\sigma^2)^2}}\, d\bar{w}_t^2$$

$$= \frac{\sigma^2\gamma^2}{\sqrt{(\gamma^1\sigma^1)^2 + (\gamma^2\sigma^2)^2}}\, d\bar{w}_t^1 - \frac{\sigma^1\gamma^1}{\sqrt{(\gamma^1\sigma^1)^2 + (\gamma^2\sigma^2)^2}}\, d\bar{w}_t^2$$

$$(2.119)$$

Concluding, with relation (2.119) the system (2.115) and (2.114) can be rewritten as

$$\begin{cases} dr_t = \kappa_r[\gamma_t - r_t]\, dt + \sigma_r\, dw_t^r \\ d\gamma_t = \kappa_\gamma[\theta_\gamma - \gamma_t]\, dt + \sigma_\gamma \left[\rho\, dw_t^r + \sqrt{1 - \rho}\, dw_t^\gamma \right] \end{cases} \qquad (2.120)$$

where $\kappa_r, \kappa_\gamma, \theta_\gamma, \sigma_r, \sigma_\gamma$ are as before, while $\rho := \frac{\sigma^2\gamma^2}{\sqrt{(\gamma^1\sigma^1)^2+(\gamma^2\sigma^2)^2}}$. Furthermore, $w_t^r = \bar{w}_t^1$ and $w_t^\gamma = \bar{w}_t^2$ are independent Wiener processes, where $\bar{w}_t^1$ and $\bar{w}_t^2$ are in turn related to the original w_t^1 and w_t^2 via (2.111) and (2.116). The system (2.120) now corresponds exactly to (26) in Filipović and Trolle (2013).

We come now to the remaining affine processes in Sect. 3.2 in Filipović and Trolle (2013). One is the process ν_t, which represents the intensity of the jump process driving the default intensity of the average bank (this takes into account the default component in the actual Libor). It is defined in Filipović and Trolle (2013) through the dynamics in (29) or, more generally, in (30), and can be obtained in our setting as follows. Consider two independent factor processes of the mean-reverting square-root form in (2.6), call them Ψ_t^ν and Ψ_t^μ and identify ν_t with Ψ_t^ν and μ_t with Ψ_t^μ. Notice that, as mentioned after (2.6), the constant parameters $\bar{a}^i$ are here replaced by stochastic processes.

The last process concerns ζ_t defined in Filipović and Trolle (2013) in (32) or, more generally, in (33). It represents the liquidity risk factor in the Libor rate. Again it can be obtained in our setup by considering two further independent factor processes of the form in (2.6), call them this time Ψ_t^ζ and Ψ_t^ϵ and identify ζ_t with Ψ_t^ζ and ϵ_t with Ψ_t^ϵ.

2.7 Multiple Curve Rational Pricing Kernel Models

In this section we present a modeling approach which is different from the martingale modeling discussed earlier in the chapter. The modeling is done directly under the real-world measure P using a pricing kernel or state-price density (deflator) approach, also referred to as potential approach in Rogers (1997). We refer to the book by Hunt and Kennedy (2004) for more details on this approach and the related references. Regarding the modeling under the real-world measure, we mention also the paper by Platen and Tappe (2015) set in a single-curve HJM setup, where a so-called benchmark approach, presented in Platen and Heath (2010), is exploited. The focus in this section will be put on rational pricing models, which were introduced by Flesaker and Hughston (1996). A particularly suitable specification of these models is the one in which the bond prices and the short rate are expressed as rational functions of one or a linear combination of several Markovian factors—this is why these models are referred to as linear rational models. The factors are often given a lognormal dynamics in which case the model is called rational lognormal model. The appealing features of linear rational models is that they can easily ensure positive interest rates and they allow for analytic pricing formulas for caps and swaptions; see in particular Filipović et al. (2014) for a detailed analysis of the properties of these models. Regarding further models in which the bond prices and interest rates are expressed as functionals of Markovian factors we refer also to Hunt et al. (2000), and among papers studying the pricing kernel approach, we mention Brody and Hughston (2004) (pricing kernels based on the Wiener chaos expansion technique), Hughston and Macrina (2012) and Akahori et al. (2014) (heat kernel approach for pricing kernels, also referred to as information-based approach), as well as other related studies cited in these papers.

The extensions of the rational pricing kernel approach to multiple curves have very recently appeared in Crépey et al. (2015b) and Nguyen and Seifried (2015). In particular, Nguyen and Seifried (2015) exploit the foreign exchange analogy to construct their model using multiplicative spreads and Crépey et al. (2015b) provide also the CVA computations in addition to the clean pricing model based on the direct modeling of the forward Libor rate. As calibration examples, both papers study a two-factor lognormal model specification, which is shown to provide very good fit to swaption market data. Note that these models allow to establish a link between a risk-neutral measure and the real-world measure—an important feature for analyzing and managing risk exposures required to comply with regulatory obligations, especially in insurance.

Below we present the main ideas of the construction of multiple curve rational pricing kernel models.

Let a probability space $(\Omega, \mathscr{F}, (\mathscr{F}_t)_{0 \leq t \leq T^*}, P)$ be given, where T^* is a finite time horizon, the filtration $(\mathscr{F}_t)_{0 \leq t \leq T^*}$ is assumed to satisfy the usual conditions and P is the real-world probability measure. The pricing kernel $(\pi_t)_{0 \leq t \leq T^*}$ is a semimartingale such that for a given generic, non-dividend paying asset with cash flow S_T at a fixed date T, its arbitrage-free price at time $t \leq T$, denoted by S_{tT}, can be expressed as

$$S_{tT} = \frac{1}{\pi_t} E^P\{\pi_T S_T | \mathscr{F}_t\} \tag{2.121}$$

For a detailed account on the pricing kernel approach we refer to Hunt and Kennedy (2004). Note that as soon as the pricing kernel π and the cash flow S_T are specified, the above expression provides a pricing formula for the asset S at all times t. In order to apply now this setup to the multiple curves, one considers the OIS bonds and the FRA contracts as the basic assets.

More precisely, denoting as earlier by $p(t, T)$ the OIS bond price at time t, Eq. (2.121) yields

$$p(t, T) = \frac{1}{\pi_t} E^P\{\pi_T | \mathscr{F}_t\} \tag{2.122}$$

since the cash flow at maturity T is $p(T, T) = 1$. In general, it can be shown that the term structure $T \mapsto p(t, T)$ is decreasing in T (i.e. the implied interest rates are nonnegative) if and only if the pricing kernel π is a nonnegative supermartingale, see e.g. Proposition 2.1 in Nguyen and Seifried (2015).

Next consider the FRA contracts. Recall that an FRA, as given in Definition 1.3, with the inception date T, maturity $T + \Delta$, fixed rate R and notional N, provides the following cash flow at maturity $T + \Delta$

$$H_{T+\Delta} = N\Delta(L(T; T, T + \Delta) - R)$$

Applying again the general pricing equation (2.121) results in the price process for the FRA given by

$$H_{t,T+\Delta} = \frac{1}{\pi_t} E^P\{\pi_{T+\Delta} N \Delta(L(T; T, T + \Delta) - R) | \mathscr{F}_t\}$$

Thus, the fair FRA rate R_t at time t is

$$R_t = \frac{\frac{1}{\pi_t} E^P\{\pi_{T+\Delta} L(T; T, T + \Delta) | \mathscr{F}_t\}}{\frac{1}{\pi_t} E^P\{\pi_{T+\Delta} | \mathscr{F}_t\}}$$

$$= \frac{1}{p(t, T + \Delta)} \frac{1}{\pi_t} E^P\{\pi_{T+\Delta} L(T; T, T + \Delta) | \mathscr{F}_t\} \tag{2.123}$$

Based on this expression Crépey et al. (2015b) define the Libor process $\tilde{L}(t; T, T + \Delta)$ by

$$\tilde{L}(t; T, T + \Delta) := \frac{1}{\pi_t} E^P \{\pi_{T+\Delta} \tilde{L}(T; T, T + \Delta) | \mathscr{F}_t\} \qquad (2.124)$$

putting $\tilde{L}(T; T, T + \Delta) := L(T; T, T + \Delta)$. Hence, the FRA rate can be written as

$$R_t = \frac{\tilde{L}(t; T, T + \Delta)}{p(t, T + \Delta)} \qquad (2.125)$$

Note that the Libor process is not exactly the forward Libor rate as given in Definition 1.2, but rather the forward Libor rate multiplied by the OIS discount factor $p(t, T + \Delta)$ for the corresponding maturity. We point out that Crépey et al. (2015b) use the notation $L(t; T, T + \Delta)$ to denote the Libor process, which in this section we have replaced with $\tilde{L}(t; T, T + \Delta)$ to distinguish it from the notation $L(t; T, T + \Delta)$ used in this monograph for the forward Libor rate. Nguyen and Seifried (2015) write $H_{t,T+\Delta}$ as

$$H_{t,T+\Delta} = \frac{1}{\pi_t} E^P \{\pi_T p(T, T + \Delta) N \Delta (L(T; T, T + \Delta) - R) | \mathscr{F}_t\}$$

$$= \frac{1}{\pi_t} E^P \left\{ \pi_T p(T, T + \Delta) N \left(\frac{\Sigma(T; T, T + \Delta)}{p(T, T + \Delta)} - (1 + \Delta R) \right) \middle| \mathscr{F}_t \right\}$$

$$= \frac{1}{\pi_t} E^P \{\pi_T N (\Sigma(T; T, T + \Delta) - p(T, T + \Delta)(1 + \Delta R)) | \mathscr{F}_t\} \quad (2.126)$$

which follows by expressing the $\mathscr{F}_T$-measurable cash flow $H_{T+\Delta}$ at time $T + \Delta$ as a cash flow $p(T, T + \Delta)H_{T+\Delta}$ at time T using the OIS discounting. Again we have adjusted the notation used in Nguyen and Seifried (2015) to correspond to the notation used in this monograph, in particular instead of $S_m^\Delta(T, T)$ therein we use $\Sigma(T; T, T + \Delta)$ to denote the multiplicative spread as defined in (1.34). Note moreover that π_t corresponds exactly to the state-price density D_t in Nguyen and Seifried (2015). Introducing now a process π^Δ such that $\pi_t^\Delta = \pi_t \Sigma(t; t, t + \Delta)$, the fair FRA rate R_t, and thus also the forward Libor rate $L(t; T, T + \Delta)$, is given by

$$L(t; T, T + \Delta) = R_t = \frac{1}{\Delta} \left(\frac{p(t, T)}{p(t, T + \Delta)} \frac{E^P \{\pi_T^\Delta | \mathscr{F}_t\}}{E^P \{\pi_T | \mathscr{F}_t\}} - 1 \right) \qquad (2.127)$$

Summarizing, Eq. (2.124) together with (2.125) and Eq. (2.127) establish a link between the forward Libor rate, denoted again by $L(t; T, T + \Delta)$, and the pricing kernel π via the OIS bonds. In Crépey et al. (2015b) this means that in order to construct the rational pricing kernel model one has to specify the spot Libor rate $L(T; T, T + \Delta)$ and in Nguyen and Seifried (2015) it means specifying the process π^Δ in addition to π. Both papers suggest a linear rational pricing model, in which the OIS bond prices $p(t, T)$ and the forward Libor rates $L(t; T, T + \Delta)$ are given as rational functions of linear combinations of several Markovian factors. Skipping the details, whereby referring the interested reader to both papers, one obtains that the OIS bond price $p(t, T)$ is given by

$$p(t, T) = \frac{p(0, T) + b^1(T)M_t^1}{p(0, t) + b^1(t)M_t^1} \tag{2.128}$$

where $t \mapsto p(0, t)$ is the initial term structure of the OIS bonds, b^1 is a deterministic function and M^1 a Markovian factor, which is a positive martingale with $M_0^1 = 0$ such that the process $p(0, t) + b^1(t)M_t^1$ is a nonnegative supermartingale with respect to P. The forward Libor rate is given by

$$L(t; T, T + \Delta) = \frac{L(0; T, T + \Delta) + b^2(T, T + \Delta)M_t^2 + b^3(T, T + \Delta)M_t^3}{p(0, T + \Delta) + b^1(T + \Delta)M_t^1} \tag{2.129}$$

where b^2 and b^3 are deterministic functions and M^2 and M^3 are Markovian factors, which are martingales with respect to P. Moreover, from (2.128) it easily follows that the short rate $r_t = -\frac{\partial}{\partial T} \log p(t, T)|_{T=t}$ is given by

$$r_t = -\frac{\frac{d}{dt}p(0, t) + \frac{d}{dt}\left(b^1(t)\right)M_t^1}{p(0, t) + b^1(t)M_t^1}$$

We emphasize that the approach presented in Crépey et al. (2015b) is slightly different, using an auxiliary probability measure, equivalent to P, and assuming that the pricing kernel π is given under this measure. Then Eq. (2.122) and the corresponding measure change are used to infer the OIS bond prices and the forward Libor rates. Therefore, the model presented above is only a special case of the more general approach in Crépey et al. (2015b). Note that our Markovian factors M^i, $i = 1, 2, 3$, are martingales with respect to the real-world measure P, whereby the Markovian factors denoted A^i, $i = 1, 2, 3$, in their paper, are martingales with respect to the auxiliary probability measure. On the other hand, Nguyen and Seifried (2015) define the pricing kernel π and the process π^Δ as linear functions of the Markovian factors directly under the real-world measure P.

Remark 2.12 The framework developed in Crépey et al. (2015b), where the model is set up under an auxiliary probability measure, linked in a suitable way to the real-world measure P, allows for quite substantial modeling freedom. In particular, the joint dynamics of the factors under P may in general be quite complicated, capturing some desirable statistical properties (for instance dependence, or a known law at some fixed future time point). However, one may then begin by using an auxiliary measure denoted by M, equivalent to P, with respect to which the factors may be independent or have a joint law which allows for semi-closed-form pricing formulas. Thus, one can do all the calculations thanks to the tractability of the model under the measure M, while achieving the desired statistical properties under the real-world measure P. This is particularly important in view of risk management which is done under the real-world measure. This line of reasoning can be extended to the risk-neutral measure Q, as pointed out in Sect. 2.3 of Crépey et al. (2015b). The model presented above can thus also be developed under the risk-neutral measure Q, starting by specifying the short rate r as it was done in the short-rate models presented earlier in this chapter. More precisely, the short rate is constructed as a rational function of the Markovian factor process M_t^1 under the auxiliary measure M, together with the density process between this measure and the risk-neutral measure Q. The explicit link between the measures Q and M allows in particular to calibrate the model to option price data under Q, using the tractable pricing formulas under M. In this sense the pricing kernel approach can be seen as a real-world measure modeling framework which is a counterpart of the martingale modeling short-rate approach, as they are both based on the bottom-up modeling idea.

Remark 2.13 (Lognormal driving factors) The martingales M^i, $i = 1, 2, 3$, in (2.128) and (2.129) can be specified in various ways. Often they are assumed to be of the form $M_t^i = M_i(t, X_t^i)$, where X^i are Markov processes under P and M_i some deterministic functions. The examples considered in Crépey et al. (2015b) and Nguyen and Seifried (2015) are lognormal, i.e.

$$M_t^i = \exp\left(a_i w_t^i - \frac{1}{2}a_i^2 t\right) - 1 \tag{2.130}$$

where w^i are Wiener processes and $a_i > 0$ are the volatility parameters of the factors, for $i = 1, 2, 3$. In particular, Crépey et al. (2015b) assume $w^1 = w^3$ and w^1 and w^2 are correlated with correlation parameter ρ. Nguyen and Seifried (2015) consider a two-factor model driven by M^1 and M^2, given as above, for both the OIS bond prices and the forward Libor rates, while allowing the volatilities a_i to be time-dependent.

Remark 2.14 (Pricing and calibration) Analogously to the previous parts of the chapter, the next step is to derive the pricing formulas for linear and optional derivatives in this setup. As shown in both Crépey et al. (2015b) and Nguyen and Seifried (2015), the pricing formulas for linear derivatives are easily obtained by combining the general formulas from Sect. 1.4, part A, with (2.128) and (2.129). For caps and swaptions the pricing formulas are obtained by using the general pricing formulas in Sect. 1.4, part B, Eqs. (2.128) and (2.129), as well as the log-normal distributions of the factors, which allows to obtain semi-explicit (up to the numerical root finding for the payoff function) pricing formulas as integrals over the Gaussian densities. We refer to both papers for the explicit expressions in all cases. We emphasize, however, that thanks to the rational structure of the model, the pricing of swaptions is especially convenient. In particular, the payoff function contains a sum of only three exponential terms related to the two Markovian factors, hence the numerical computation of the price is a very easy task. The swaption price formulas can be found in Sect. 3.3 in Crépey et al. (2015b) and Sect. 5.2 in Nguyen and Seifried (2015).

Concerning calibration, the examples provided in both papers show that a two-factor lognormal rational model, despite the simple dynamics of the driving factors in (2.130), provides a very good fit to the swaption market prices without need for stochastic volatility or more general Lévy drivers. This property allows to profit from the simple dynamics of the underlying processes, which are easy to simulate in an exact and fast manner.

Chapter 3
Multiple Curve Heath–Jarrow–Morton (HJM) Framework

This chapter concerns the HJM framework in a multi-curve setup and is mainly based on the papers Fujii et al. (2011), Crépey et al. (2012), Moreni and Pallavicini (2014), Crépey et al. (2015a), and the Ph.D. thesis by Miglietta (2015). As in Chap. 2, also in this chapter we shall model a basic OIS rate and the various risky rates as spreads over the OIS rate. While the rates modeled in Chap. 2 are short rates and short-rate spreads, here we deal with forward rates and their spreads. Since the actual risky rates are generally larger than the corresponding OIS rates, this can be obtained by modeling the additive spreads possibly as positive processes.

The HJM framework is situated in between the pure "bottom-up" short-rate framework and the pure "top-down" framework of the LMMs. With respect to the short-rate framework, it has the advantage of allowing for an automatic fit to the initial Libor and OIS term structures (as well as for convenient calibration of the model parameters to derivative prices). With respect to the LMMs, depending on the volatility structure, it allows for low-dimensional Markovian representations of the term structure and this turns out to be beneficial in various contexts, in particular in CVA and other valuation adjustment computations. In addition to the pure HJM approach, in this chapter we shall also consider an approach that can be seen as a hybrid HJM-LMM approach and which can be advantageous in the multiple curve setup.

Since the ultimate goal of interest rate theory is the pricing of interest rate derivatives and the main underlying quantity in these derivatives are the Libor rates, one of the first objectives is to derive models for the dynamics of the Libor rates that are arbitrage-free. For simplicity of exposition, and analogously to Moreni and Pallavicini (2014) and Miglietta (2015), we shall limit ourselves to Wiener driven models, while Crépey et al. (2012) and Crépey et al. (2015a) consider Lévy driven models although with deterministic coefficients. By allowing volatility to be stochastic, one may still account for smile effects and thereby compensate for possible shortcomings of Wiener modeling with respect to Lévy modeling in case of deterministic coefficients.

In order to develop an arbitrage-free model in the multiple curve HJM setup, we shall in the first step consider a model for the OIS prices, set up directly under the

© The Author(s) 2015
Z. Grbac and W.J. Runggaldier, *Interest Rate Modeling: Post-Crisis Challenges and Approaches*, SpringerBriefs in Quantitative Finance,
DOI 10.1007/978-3-319-25385-5_3

martingale measure Q. This model is developed in complete analogy to the standard pre-crisis HJM model. In the second step, suitable quantities connected to the FRA contracts will be chosen and also modeled in the spirit of the HJM framework in such a way that the complete model is free of arbitrage. To identify those quantities, we first recall the expressions for the spot and forward Libor rates as given by (see Remark 1.4 with (2.33), as well as (1.18))

$$
\begin{aligned}
L(T; T, T + \Delta) &= \tfrac{1}{\Delta} \left(\tfrac{1}{\bar{p}(T, T+\Delta)} - 1 \right) \\
L(t; T, T + \Delta) &= E^{T+\Delta} \{ L(T; T, T + \Delta) \mid \mathscr{F}_t \} \\
&= \tfrac{1}{\Delta} E^{T+\Delta} \left\{ \left(\tfrac{1}{\bar{p}(T, T+\Delta)} - 1 \right) \Big| \mathscr{F}_t \right\}
\end{aligned}
\tag{3.1}
$$

where, we recall, the symbol $\bar{p}(t, T)$ was introduced to denote generically fictitious risky bond prices that are assumed to be affected by the same factors as the Libor rates (*can be seen as a bond issued by a representative bank from the Libor group*) and are such that the classical relationship between forward rates and bonds is preserved also for the actual Libor rates, but only at the spot level (first equation above).

Since the HJM framework is situated in between the short-rate models and the LMMs, the following two possible approaches may be considered:

(i) Mimicking the LMMs, model directly the forward Libor rates in a suitable way thereby establishing also a link with Chap. 4. We shall call this the hybrid HJM-LMM approach. We shall do this in Sect. 3.2, where we consider the following quantities, namely the forward Libor rates multiplied by the length of the interval to which they apply

$$
G(t; T, T + \Delta) = \Delta E^{T+\Delta} \{ L(T; T, T + \Delta) \mid \mathscr{F}_t \}
\tag{3.2}
$$

that, by definition, are martingales under the forward measure. The no-arbitrage conditions, therefore, translate directly into martingale conditions on the process $G(\cdot; T, T + \Delta)$. Another possibility, which is more in the spirit of the HJM approach, is to introduce the fictitious bonds as in (3.1), but to require that the relationship between the forward rates and the bonds is preserved, not only at the spot level, but at all levels t. Introducing a new notation $p^{\Delta}(t, T)$ to emphasize this difference, we therefore have

$$
\begin{aligned}
L(t; T, T + \Delta) &= E^{T+\Delta} \{ L(T; T, T + \Delta) \mid \mathscr{F}_t \} \\
&= \tfrac{1}{\Delta} E^{T+\Delta} \left\{ \left(\tfrac{1}{p^{\Delta}(T, T+\Delta)} - 1 \right) \Big| \mathscr{F}_t \right\} =: \tfrac{1}{\Delta} \left(\tfrac{p^{\Delta}(t, T)}{p^{\Delta}(t, T+\Delta)} - 1 \right)
\end{aligned}
\tag{3.3}
$$

The modeling quantities are thus the $p^{\Delta}(t, T)$-bonds, whose dynamics are modeled according to an HJM approach. The no-arbitrage conditions, hence, translate into martingale conditions on the ratios $\frac{p^{\Delta}(\cdot, T)}{p^{\Delta}(\cdot, T+\Delta)}$ of fictitious bond prices.

(ii) The second HJM approach is to obtain, as in Chap. 2, the dynamics of $L(\cdot; T, T + \Delta)$ via dynamics of $\bar{p}(\cdot, T)$, where the connection between the Libor rates and the $\bar{p}(\cdot, T)$-bonds is established only at the spot level. The difference with Chap. 2

is the way in which the dynamics of $\bar{p}(\cdot, T)$ are modeled. To this effect we shall consider two specific ways to interpret these bond prices, which are inspired by a credit risk analogy and a foreign exchange analogy, both situated in the HJM setup. Regarding $\bar{p}(\cdot, T)$-bonds, each of these two interpretations will yield specific conditions on their dynamics to exclude arbitrage possibilities in the respective models. In line with the traditional HJM approach, we shall also consider the instantaneous forward rates $\bar{f}(t, T) := -\frac{\partial}{\partial T} \log \bar{p}(t, T)$, associated to $\bar{p}(t, T)$ and assume that they are given by

$$\bar{f}(t, T) = f(t, T) + g(t, T)$$

where $f(t, T)$ are the forward rates corresponding to the OIS bonds $p(t, T)$, while $g(t, T)$ are the forward rate spreads. This approach will be the subject of Sect. 3.3. In these models the crucial quantity for computations of derivative prices is given by the following conditional expectation (see also (2.50))

$$\bar{\nu}_{t,T} = E^{T+\Delta} \left\{ \frac{1}{\bar{p}(T, T + \Delta)} \Big| \mathscr{F}_t \right\} \tag{3.4}$$

Summarizing the definitions introduced above, we get the following equivalent representations for the forward Libor rate

$$L(t; T, T + \Delta) = \begin{cases} \frac{1}{\Delta} G(t; T, T + \Delta) & \text{(used in Section 3.2)} \\ \frac{1}{\Delta} \left(\frac{p^\Delta(t, T)}{p^\Delta(t, T + \Delta)} - 1 \right) & \text{(used in Section 3.2)} \\ \frac{1}{\Delta} \left(\bar{\nu}_{t,T} - 1 \right) & \text{(used in Section 3.3)} \end{cases} \tag{3.5}$$

from which it also follows that

$$\bar{\nu}_{t,T} = \frac{p^\Delta(t, T)}{p^\Delta(t, T + \Delta)} = G(t; T, T + \Delta) + 1 \tag{3.6}$$

and these relations will be useful in the pricing of linear interest rate derivatives. Notice that the chosen ordering of (i) and (ii) reflects the fact that the models in (i) lead to intrinsic no-arbitrage conditions in reference to the traded assets (FRA contracts), while the models in (ii) lead to what one may call pseudo no-arbitrage conditions since they stem from interpreting the fictitious bonds in a specific way, rather than from absence of arbitrage with respect to the traded FRA contracts, which is ensured already by model construction (note the martingale property of $\bar{\nu}_{t,T}$ resulting from (3.4)).

Regarding work from the literature mentioned at the beginning of this introduction, Fujii et al. (2011), Moreni and Pallavicini (2014), Crépey et al. (2015a) and Miglietta (2015) focus on the first HJM-LMM hybrid approach, whereas Crépey et al. (2012) is based on the credit risk analogy and Miglietta (2015) briefly mentions a possible foreign exchange analogy, however, without providing the details for the HJM setup. The authors in Moreni and Pallavicini (2014) model instantaneous rates and forward Libors under the forward measure and, to do so, they use a specific representation in terms of a Markovian factor process showing the efficiency of their modeling approach for the purpose of model calibration. In the context of model calibration see furthermore Pallavicini and Tarenghi (2010), where a hybrid HJM-LMM approach is used as well. In addition to these references, we want to mention also Cuchiero et al. (2015), who study a multiplicative spread approach in a general hybrid HJM-LMM semimartingale setup, which is discussed in Sect. 4.3.

The remaining part of this chapter is structured as follows. Section 3.1 gives the necessary preliminaries on the HJM approach and its adaptation to the multiple curve setup. As already mentioned, in Sects. 3.2 and 3.3 we shall describe the modeling approaches corresponding to the above points (**i**) and (**ii**) respectively. For each of these sections we shall derive, besides the models, the no-arbitrage conditions. We shall also discuss possible volatility structures in view of obtaining Markovian representations and, as in the standard HJM, short-rate models implied by the various forward rate models. In Sect. 3.4 we shall consider derivative pricing for linear and optional interest rate derivatives in the context of all model types, those in Sect. 3.2 and those in Sect. 3.3. Finally, in Sect. 3.5 we shall carry over (when applicable) also to the HJM setting the idea of an adjustment factor as introduced in Chap. 2 (see Sect. 2.3.2).

3.1   Adapting the Classical HJM Approach

In this section we first present the model for the OIS bonds $p(\cdot, T)$ and the generic bonds related to the Libor rates that we denoted by $\bar{p}(\cdot, T)$. Later on we shall distinguish between the three possible interpretations of these bonds mentioned in the introductory part of this chapter and introduce also the corresponding notation. The HJM drift conditions for the OIS bonds, as we shall see below, will be the classical ones, whereas for the $\bar{p}(\cdot, T)$ they will depend on the specific model and the interpretation of these bonds.

We begin with the modeling part. Corresponding to the short rate r_t and the short-rate spread s_t of Chap. 2, here we start from the following quantities:

$$\begin{cases} f(t, T) & : \text{the instantaneous, continuously compounded forward OIS interest} \\ & \quad \text{rate} \\ g(t, T) & : \text{the forward rate spread} \end{cases}$$

at time t with maturity $T \geq t$, where $T \in [0, \bar{T}]$ and $\bar{T}$ is a finite time horizon. Recall that $f(t, T)$ is related to the OIS bond prices $p(t, T)$ via the usual relationship (1.12), which in turn yields

$$p(t, T) = \exp\left[-\int_t^T f(t, u)du\right] \tag{3.7}$$

For the continuously compounded forward rates $\bar{f}(t, T)$ we postulate that

$$\bar{f}(t, T) = f(t, T) + g(t, T)$$

and similarly as above we define the fictitious bond prices $\bar{p}(t, T)$ as follows (notice the analogy with (2.35))

$$\bar{p}(t, T) = \exp\left[-\int_t^T \bar{f}(t, u)du\right] = \exp\left[-\int_t^T (f(t, u) + g(t, u))du\right] \quad (3.8)$$

As in the traditional HJM approach we use martingale modeling assuming that, on $(\Omega, \mathscr{F}, (\mathscr{F}_t)_{0 \le t \le \bar{T}}, Q)$ with Q a standard martingale measure related to the OIS curve, the forward rates and spreads satisfy

$$\begin{aligned} df(t, T) &= a(t, T)dt + \sigma(t, T)\,dw_t \\ dg(t, T) &= a^*(t, T)dt + \sigma^*(t, T)\,dw_t \\ d\bar{f}(t, T) &= \bar{a}(t, T)dt + \bar{\sigma}(t, T)\,dw_t \end{aligned} \quad (3.9)$$

with $\bar{a}(\cdot) = a(\cdot) + a^*(\cdot)$, $\bar{\sigma}(\cdot) = \sigma(\cdot) + \sigma^*(\cdot)$, and where the coefficients may be generic adapted processes satisfying the implicit integrability conditions. For simplicity we assume here that the driving random process is a d-dimensional Wiener process. Note that this means that we might not necessarily have the desired positivity for the spreads. Below we shall however consider mean-reverting Gaussian models for the short-term spreads $g(t, t)$ which implies that they may become negative only with a small probability. With Lévy driven models as in Crépey et al. (2012) positivity of the spreads can be guaranteed under certain assumptions.

Letting all the required regularity assumptions, such as differentiation under the integral sign, be satisfied, for the OIS and the fictitious bond prices we obtain

$$dp(t, T) = p(t, T)\left[(r_t - A(t, T) + \frac{1}{2}|\Sigma(t, T)|^2)dt - \Sigma(t, T)dw_t\right] \quad (3.10)$$

$$d\bar{p}(t, T) = \bar{p}(t, T)\left[(\bar{r}_t - \bar{A}(t, T) + \frac{1}{2}|\bar{\Sigma}(t, T)|^2)dt - \bar{\Sigma}(t, T)dw_t\right] \quad (3.11)$$

where $A(t, T) := \int_t^T a(t, u)du$, $\Sigma(t, T) := \int_t^T \sigma(t, u)du$, $r_t = f(t, t)$ and, analogously for $\bar{A}(t, T)$, $\bar{\Sigma}(t, T)$, $\bar{r}_t$ as well as $A^*(t, T)$, $\Sigma^*(t, T)$. Recall from Sect. 1.3.2 that the discounted OIS bond prices have to be martingales under Q, which leads to the classical HJM drift condition for $A(t, T)$, and thus also for $a(t, T)$, namely

$$A(t, T) = \frac{1}{2}|\Sigma(t, T)|^2 \quad (3.12)$$

As far as the $\bar{p}(t, T)$-bonds are concerned, note that, since we are dealing with bond prices stemming from different interpretations, we cannot derive the drift conditions by requiring directly that the discounted values of $\bar{p}(t, T)$ be martingales under Q and so we proceed separately for each of the cases in the forthcoming sections.

3.2 Hybrid HJM-LMM Models

In this section we shall discuss the modeling approach outlined in (**i**) of the introductory part of this chapter, where we assume an HJM model for the OIS bond prices $p(t, T)$ and in addition we consider two possible modeling quantities related to the forward Libor rates. The first model is a model for the processes $G(\cdot; T, T + \Delta)$, which represent the Libor rates multiplied by the length of the interval to which they apply, and the second one is a model for the $p^\Delta(t, T)$-bonds.

3.2.1 HJM-LMM Model for the Forward Rates

Based on Crépey et al. (2015a), we shall model the OIS bond prices $p(t, T)$ as in Sect. 3.1 together with the processes $G(\cdot; T, T + \Delta)$ as defined in (3.2) that we recall here, namely

$$G(t; T, T + \Delta) = \Delta E^{T+\Delta} \{L(T; T, T + \Delta) \mid \mathscr{F}_t\} \qquad (3.13)$$

Notice that these processes correspond, modulo the factor Δ, to the forward Libor rates in Definition 1.2 in Chap. 1 (see also 4.2 in the next chapter) and, by definition, are martingales with respect to the corresponding forward measures. Modeling the forward Libor rates multiplied by the length of the corresponding interval Δ is a choice made for convenience reasons only; as pointed out in Crépey et al. (2015a) it gives rise to slightly simpler formulas in this set-up. These quantities are, for short maturities, directly observable on the market and, for longer maturities, can be bootstrapped from observable Libor swap rates. Mercurio and co-authors derive the dynamics of the forward Libor rates in the standard LMM framework. Here, based on Crépey et al. (2015a), we follow instead an HJM approach that for forward Libor rates appears to have been first applied by Moreni and Pallavicini (2014) and that allows to have low-dimensional Markovian representations, as well as to access the induced short-rate process r_t, which are features that are relevant for practical CVA computations. With respect to the previous chapter on short-rate models notice that, by (3.6), we model here a quantity at the level of $\bar{\nu}_{t,T}$.

3.2.1.1 Model and No-Arbitrage Conditions

As already mentioned, following Crépey et al. (2015a), we shall model here directly the dynamics of the OIS bond prices $p(t, T)$, as well as the processes $G(t; T, T + \Delta)$ in the form as they were recalled in (3.13).

   We assume given a filtered probability space $(\Omega, \mathscr{F}, (\mathscr{F}_t)_{0 \leq t \leq \bar{T}}, Q)$, where $\mathscr{F} = \mathscr{F}_{\bar{T}}$ with $\bar{T}$ a fixed finite horizon and Q is the OIS pricing measure related to B_t as a numéraire. For the OIS bond prices we start from the representation (3.10), namely

$$dp(t, T) = p(t, T) \left[(r_t - A(t, T) + \frac{1}{2} |\Sigma(t, T)|^2) dt - \Sigma(t, T) dw_t \right] \quad (3.14)$$

The coefficients have to satisfy the classical HJM drift condition (3.12). Contrary to the HJM approach for the instantaneous forward rate model of Sect. 3.1, the coefficients $A(t, T)$ and $\Sigma(t, T)$ are here given directly and do not result from an integration of the coefficients in the instantaneous forward rate dynamics.

Concerning $G(t; T, T + \Delta)$, following Crépey et al. (2015a) we impose the dynamics

$$G(t; T, T + \Delta) = G(0; T, T + \Delta) \quad (3.15)$$

$$\exp \left[\int_0^t \alpha(s; T, T + \Delta) ds + \int_0^t \sigma(s; T, T + \Delta) dw_s \right]$$

that guarantees their positivity. While in Crépey et al. (2015a) the driving process is a Lévy process and the coefficients are deterministic time functions, here we consider for simplicity a driving d-dimensional Wiener process. In exchange we may allow for stochastic volatility that partly compensates the reduction from Lévy to Wiener processes. To complete the model, it remains to impose conditions guaranteeing that $G(\cdot; T, T + \Delta)$ are, as they should be according to their definition, martingales under the corresponding forward measures. To this effect we have

Proposition 3.1 *In the model for $G(t; T, T + \Delta)$ given by (3.15) we have the no-arbitrage drift condition*

$$\alpha(t; T, T + \Delta) = -\frac{1}{2} |\sigma(t; T, T + \Delta)|^2 + \langle \sigma(t; T, T + \Delta), \Sigma(t, T + \Delta) \rangle$$

$$(3.16)$$

which, replaced in the original model (3.15), leads to an arbitrage-free model for the Libor rates under the measure Q.

Proof In order to derive the no-arbitrage condition, we first have to change the measure from Q to $Q^{T+\Delta}$. Recall from (1.14) that the density process for the change from Q to $Q^{T+\Delta}$ is given by

$$\mathscr{L}_t^{T+\Delta} := \frac{dQ^{T+\Delta}}{dQ} \bigg|_{\mathscr{F}_t} = \frac{p(t, T + \Delta)}{B_t p(0, T + \Delta)}$$

Using Itô's formula, from (3.10), the drift condition (3.12), as well as the fact that $B_t = \exp \left[\int_0^t r_s ds \right]$, we obtain

$$d\mathscr{L}_t^{T+\Delta} = -\mathscr{L}_t^{T+\Delta} \Sigma(t, T + \Delta) dw_t$$

By Girsanov's theorem it follows that $w_t^{T+\Delta}$ as given by

$$dw_t^{T+\Delta} = dw_t + \Sigma(t, T + \Delta) dt \quad (3.17)$$

is a Wiener process under $Q^{T+\Delta}$.

Again by using Itô's formula, from (3.15) together with (3.17) we obtain the following $Q^{T+\Delta}$-dynamics for $G(t; T, T + \Delta)$

$$
\begin{aligned}
d\,G(t; T, T + \Delta) = G(t; T, T + \Delta) \cdot \Big[\big(\alpha(t; T, T + \Delta) + \tfrac{1}{2} |\sigma(t; T, T + \Delta)|^2 \\
- \langle \sigma(t; T, T + \Delta), \Sigma(t, T + \Delta) \rangle \big) \, dt \\
+ \sigma(t; T, T + \Delta) \, dw_t^{T+\Delta} \Big]
\end{aligned}
$$

(3.18)

In order that $G(\cdot; T, T + \Delta)$ is a $Q^{T+\Delta}$-martingale, the drift term in (3.18) has to vanish, which implies (3.16). □

Remark 3.1 Now we can see why the model (3.14)–(3.15) presented above can be seen as a hybrid derived from the HJM and the LMM approaches. The modeling objects are the OIS bond prices as in the HJM framework and the forward Libor rates (modulo the factor Δ) as in the LMM framework. Moreover, the modeling is done under the martingale measure Q and not under the forward measures, as one would have done in the LMM spirit. The main reason for this is to profit from the low-dimensional Markovian representations that can be obtained in the HJM framework, as opposed to the LMM framework. More details on this can be found in the introduction of Crépey et al. (2015a).

3.2.1.2 Markovianity and Induced Short Rates (Vasiček-Type Volatility Structure)

In this subsection we want to show that, for specific volatility structures, which here are the OIS bond price volatility $\Sigma(t, T)$ and the volatility $\sigma(t; T, T + \Delta)$ of the process $G(\cdot; T, T + \Delta)$, one obtains a Markovian factor representation for the OIS bond prices and the process $G(\cdot; T, T + \Delta)$. Thereby one obtains also an induced dynamic model for the factors, one of which results to be the short rate r_t. We limit ourselves for illustration purposes to one of the most significant cases, namely that of a Vasiček-type term structure.

Let us begin with the OIS bond price model, which is given via the instantaneous forward rate $f(t, T)$, whose dynamics is

$$
f(t, T) = f(0, T) + \int_0^t \langle \Sigma(s, T), \sigma(s, T) \rangle ds + \int_0^t \sigma(s, T) dw_s
$$

(3.19)

which follows from (3.9) by using the drift condition (3.12) for $a(t, T)$ in terms of $\sigma(t, T)$. This representation is infinite-dimensional and, in complete analogy to the classical HJM, we may investigate whether with a suitable choice of the volatilities $\sigma(t, T)$ we may obtain a finitely parametrized representation and a Markovian model for the induced short rate $r_t = f(t, t)$.

We start from a two-dimensional driving $(Q, \mathscr{F}_t)$-Wiener process $w_t = (w_t^1, w_t^2)$ and, for given positive constants σ, σ^*, as well as non-zero $b, b^* \in \mathbb{R}$, we let the volatilities be given by

$$\sigma(t, T) \qquad = \left(\sigma e^{-b(T-t)}, 0\right)$$

$$\sigma(t; T, T + \Delta) = \left(\tfrac{\sigma}{b} e^{-b(T-t)} \left(1 - e^{-b\Delta}\right), \tfrac{\sigma^*}{b^*} e^{-b^*(T-t)} \left(1 - e^{-b^*\Delta}\right)\right) \tag{3.20}$$

respectively, where $\Sigma(t, T) = \int_t^T \sigma(t, u)du$.

We consider first the forward rates $f(t, T)$ given above and the induced short rate r_t. From (3.20) we obtain

$$
\begin{aligned}
f(t, T) &= f(0, T) + \int_0^t \tfrac{\sigma^2}{b} \left(e^{-b(T-s)} - e^{-2b(T-s)}\right) ds + \sigma e^{-bT} \int_0^t e^{bs} dw_s^1 \\
&= f(0, T) + \tfrac{\sigma^2}{b^2} e^{-bT} \left(e^{bt} - 1 - \tfrac{1}{2}e^{-bT+2bt} + \tfrac{1}{2}e^{-bT}\right) + \sigma e^{-bT} \int_0^t e^{bs} dw_s^1 \\
&= f(0, T) + \tfrac{\sigma^2}{b^2} \left(e^{-b(T-t)} - e^{-bT}\right) \\
&\quad + \tfrac{\sigma^2}{2b^2} \left(e^{-2bT} - e^{-2b(T-t)}\right) + \sigma e^{-bT} \int_0^t e^{bs} dw_s^1
\end{aligned}
\tag{3.21}
$$

It follows that the short rate $r_t = f(t, t)$ is given by

$$
\begin{aligned}
r_t &= f(0, t) + \tfrac{\sigma^2}{b^2} \left(1 - e^{-bt}\right) + \tfrac{\sigma^2}{2b^2} \left(e^{-2bt} - 1\right) + \sigma e^{-bt} \int_0^t e^{bs} dw_s^1 \\
&=: m(t) + \sigma e^{-bt} \int_0^t e^{bs} dw_s^1
\end{aligned}
\tag{3.22}
$$

This implies

$$
\begin{aligned}
dr_t &= \left(m'(t) - b\sigma e^{-bt} \int_0^t e^{bs} dw_s^1\right) dt + \sigma \, dw_t^1 \\
&= \left(m'(t) - b(r_t - m(t))\right)dt + \sigma \, dw_t^1
\end{aligned}
\tag{3.23}
$$

Defining

$$
\begin{aligned}
\rho(t) &:= m(t) + \tfrac{1}{b} m'(t) \\
&= f(0, t) + \tfrac{1}{b} \tfrac{\partial}{\partial t} f(0, t) + \tfrac{\sigma^2}{2b^2} \left(1 - e^{-2bt}\right)
\end{aligned}
\tag{3.24}
$$

we obtain that the induced short rate r_t satisfies the following Hull–White extended Vasiček model

$$dr_t = b(\rho(t) - r_t)dt + \sigma \, dw_t^1 \tag{3.25}$$

Since we have followed an HJM approach by which the model is automatically calibrated to the initial term structure $f(0, t)$, at least one of the coefficients in the Vasiček-type model (3.25), namely $\rho(t)$, is infinite-dimensional (*Hull–White extension*).

Next we try to see whether in the present setting, where we model directly the process $G(\cdot; T, T + \Delta)$, we are able to obtain quantities that can be viewed as corresponding to a spread (forward or short-rate spread). We begin by deriving an explicit expression for $G(t; T, T + \Delta)$, thereby using the shorthand notation $G(t, T)$ for $G(t; T, T + \Delta)$ and, similarly, with $\alpha(t, T)$. The starting point is (3.15), where $\alpha(t, T)$ has to be chosen so that the no-arbitrage drift condition (3.16) is satisfied. Given (3.20), this drift condition allows one to obtain an explicit expression for $\alpha(t, T)$ as a function of $(b, b^*, \sigma, \sigma^*)$; for simplicity of notation, below we shall keep this symbol $\alpha(t, T)$ to represent the above function of $(b, b^*, \sigma, \sigma^*)$. Furthermore, we shall also use the function $m(t)$, implicitly defined in (3.22) so that, always by (3.22), we may write

$$\sigma e^{-bt} \int_0^t e^{bs} dw_s^1 = r_t - m(t)$$

Then we first have that the OIS bond price $p(t, T)$ can be expressed as an exponentially affine function of the current level of the short rate r_t. More precisely, inserting the volatility (3.20) into the bond price expression (3.56) we have

$$p(t, T) = \exp[a(t, T) + b(t, T)r_t]$$

where

$$a(t, T) = \log\left(\frac{p(0, T)}{p(0, t)}\right) + \int_0^t (A(s, t) - A(s, T))\, ds - b(t, T)m(t)$$

and

$$b(t, T) = \frac{1}{b}\left(e^{-b(T-t)} - 1\right)$$

as in the classical HJM setup with Vasiček volatility. Moreover, for $G(t, T)$ we have

$$G(t, T) = G(0, T)\,\exp\Big[\int_0^t \alpha(s, T)ds + \int_0^t \tfrac{\sigma}{b} e^{-b(T-s)}\left(1 - e^{-b\Delta}\right) dw_s^1$$
$$+ \int_0^t \tfrac{\sigma^*}{b^*} e^{-b^*(T-s)}\left(1 - e^{-b^*\Delta}\right) dw_s^2\Big]$$

$$= G(0, T)\,\exp\Big[\int_0^t \alpha(s, T)ds + \tfrac{\sigma}{b} e^{bt}\left(e^{-bT} - e^{-b(T+\Delta)}\right) e^{-bt}\int_0^t e^{bs} dw_s^1$$
$$+ \tfrac{\sigma^*}{b^*} e^{b^*t}\left(e^{-b^*T} - e^{-b^*(T+\Delta)}\right) e^{-b^*t}\int_0^t e^{b^*s} dw_s^2\Big]$$

$$= G(0, T)\,\exp\Big[\int_0^t \alpha(s, T)ds + \tfrac{1}{b} e^{bt}\left(e^{-bT} - e^{-b(T+\Delta)}\right) (r_t - m(t))$$
$$+ \tfrac{\sigma^*}{b^*} e^{b^*t}\left(e^{-b^*T} - e^{-*b(T+\Delta)}\right) e^{-b^*t}\int_0^t e^{b^*s} dw_s^2\Big]$$

$$= \exp[m(t, T) + n(t, T)r_t + n^*(t, T)q_t]$$

$$(3.26)$$

where

$$m(t, T) = \log(G(0, T)) + \int_0^t \alpha(s, T)ds - \frac{1}{b} e^{bt}\left(e^{-bT} - e^{-b(T+\Delta)}\right) m(t)$$

and is thus completely specified by $G(0, T)$ and $(b, b^*, \sigma, \sigma^*)$. Furthermore,

$$\begin{cases} n(t,T) = \frac{1}{b}e^{bt}\left(e^{-bT} - e^{-b(T+\Delta)}\right) \\ n^*(t,T) = \frac{\sigma^*}{b^*}e^{b^*t}\left(e^{-b^*T} - e^{-b^*(T+\Delta)}\right) \end{cases}$$

which are also completely specified in terms of $(b, b^*, \sigma, \sigma^*)$. Finally

$$q_t := e^{-b^*t}\int_0^t e^{b^*s}dw_s^2 \tag{3.27}$$

and satisfies thus

$$dq_t = -b^*q_t dt + dw_t^2, \quad q_0 = 0 \tag{3.28}$$

Notice that all dynamics are under the standard martingale measure Q.

Remark 3.2 By (3.26), we have ended up with a two-factor exponentially affine Markovian representation for $G(t; T, T + \Delta)$. Of the two Markovian factors that drive $G(t; T, T + \Delta)$, one is the short rate r_t. The other factor would in a sense correspond to the short-rate spread considered in Chap. 2 (see also Sect. 3.2.2.2, where the second factor is s_t that satisfies (3.43) and corresponds to a short-rate spread). Here it has however no specific economic meaning. In fact, by deriving the induced short-rate model, we follow a top-down approach and so not all resulting quantities at the lower level need to have a clear economic meaning. The situation is different when following instead a bottom-up approach as in Chap. 2, where we start from the quantities at the lowest (lower) level.

3.2.2 HJM-LMM Model for the Fictitious Bond Prices

We recall that we shall use the symbol $p^\Delta(t, T)$ for the fictitious bond prices that we sometimes refer to also as Libor bonds, which are defined by the following implicit relation with the forward Libor rates (see (3.1) and (3.3))

$$\begin{aligned} L(T; T, T + \Delta) &= \frac{1}{\Delta}\left(\frac{1}{p^\Delta(T,T+\Delta)} - 1\right) \\ L(t; T, T + \Delta) &= \frac{1}{\Delta}E^{T+\Delta}\left\{\left(\frac{1}{p^\Delta(T,T+\Delta)} - 1\right)\Big|\mathcal{F}_t\right\} = \frac{1}{\Delta}\left(\frac{p^\Delta(t,T)}{p^\Delta(t,T+\Delta)} - 1\right) \end{aligned} \tag{3.29}$$

In words, the fictitious bonds $p^\Delta(\cdot, T + \Delta)$ are bonds consistent with the forward Libor curve $L(\cdot; T, T + \Delta)$ in the sense of the second relation in (3.29). This is done in order to reproduce the classical relationship between the Libor rates and the bond prices not only at the level of the spot Libor rates $L(T; T, T + \Delta)$; compare Eq. (1.11) in Sect. 1.3 where the same was done for the initial Libor curve at time $t = 0$. Note that here the definition of the bond prices $p^\Delta(\cdot, T + \Delta)$ implies the martingale condition on the ratio $\frac{p^\Delta(t,T)}{p^\Delta(t,T+\Delta)}$ with respect to the forward measure $Q^{T+\Delta}$. These fictitious bonds in the form of $p^\Delta(t, T)$ were considered also in Kenyon (2010) and correspond to what in Henrard (2014) are called *pseudo-discount factor curves* (see Chap. 3 there). Henrard (2014) also sees the relationship (3.29) as being of the "wrong number used in the wrong formula to obtain the correct result" type of approach.

3.2.2.1 Model and No-Arbitrage Condition for the $p^\Delta(t, T)$-Bonds

Recall that for the $p^\Delta(t, T)$-bonds we had assumed (3.29), where $L(t; T, T + \Delta)$ is a $Q^{T+\Delta}$-martingale, implying that also the process $\nu^\Delta_{t,T} := \frac{p^\Delta(t,T)}{p^\Delta(t,T+\Delta)}$ is a $Q^{T+\Delta}$-martingale. This martingale condition is directly implied by the absence of arbitrage imposed on the FRA contracts. As usual, the no-arbitrage condition can be derived by determining the drift in the dynamics of $\nu^\Delta_{t,T}$ under $Q^{T+\Delta}$ and setting it equal to zero.

The model for the dynamics of the OIS bond prices $p(t, T)$ and the fictitious bond prices $p^\Delta(t, T)$ is obtained following the HJM approach outlined in Sect. 3.1, which is based on the modeling of the instantaneous forward rate $f(t, T)$ and the forward spread $g(t, T)$, which are given by their dynamics in (3.9), namely

$$df(t, T) = a(t, T)dt + \sigma(t, T) dw_t$$
$$dg(t, T) = a^*(t, T)dt + \sigma^*(t, T) dw_t \tag{3.30}$$

As in Sect. 3.1, the dynamics of the OIS bond prices is thus given by

$$dp(t, T) = p(t, T)\left[(r_t - A(t, T) + \frac{1}{2}|\Sigma(t, T)|^2)dt - \Sigma(t, T)dw_t\right] \tag{3.31}$$

with the drift condition (3.12). Replacing all the quantities denoted by a "bar" in Sect. 3.1 with a superscript Δ, we obtain the following dynamics for the $p^\Delta(t, T)$-bond prices

$$dp^\Delta(t, T) = p^\Delta(t, T)\left[(r_t^\Delta - A^\Delta(t, T) + \frac{1}{2}|\Sigma^\Delta(t, T)|^2)dt - \Sigma^\Delta(t, T)dw_t\right] \tag{3.32}$$

We have now

Proposition 3.2 *For the model (3.32), we have the following no-arbitrage drift condition*

$$A^\Delta(t, T + \Delta) - A^\Delta(t, T) = -\frac{1}{2}|\Sigma^\Delta(t, T + \Delta)) - \Sigma^\Delta(t, T)|^2$$
$$+ \langle \Sigma(t, T + \Delta), \Sigma^\Delta(t, T + \Delta) - \Sigma^\Delta(t, T)\rangle \tag{3.33}$$

Proof On the basis of (3.32) we have

$$
\begin{aligned}
dv_{t,T}^{\Delta} &= v_{t,T}^{\Delta}\Big[(A^{\Delta}(t,T+\Delta) - A^{\Delta}(t,T) + \tfrac{1}{2}|\Sigma^{\Delta}(t,T+\Delta)) - \Sigma^{\Delta}(t,T|^2)dt \\
&\quad + (\Sigma^{\Delta}(t,T+\Delta) - \Sigma^{\Delta}(t,T))dw_t\Big] \\
&= v_{t,T}^{\Delta}\Big[(A^{\Delta}(t,T+\Delta) - A^{\Delta}(t,T) + \tfrac{1}{2}|\Sigma^{\Delta}(t,T+\Delta)) - \Sigma^{\Delta}(t,T)|^2 \\
&\quad - \langle \Sigma(t,T+\Delta), \Sigma^{\Delta}(t,T+\Delta) - \Sigma^{\Delta}(t,T)\rangle dt \\
&\quad + (\Sigma^{\Delta}(t,T+\Delta) - \Sigma^{\Delta}(t,T))dw_t^{T+\Delta}\Big]
\end{aligned}
$$

$$(3.34)$$

where $w_t^{T+\Delta}$ is a Wiener process under $Q^{T+\Delta}$ as given in (3.17) and so the condition (3.33) follows immediately.   □

Remark 3.3 Contrary to the standard HJM drift condition, condition (3.33) does not define uniquely the coefficient $A^{\Delta}(t,T)$; in fact, we get a constraint only on $A^{\Delta}(t,T+\Delta) - A^{\Delta}(t,T)$. This is related to the non-unique definition of $p^{\Delta}(t,T)$ via the second relation in (3.29) in terms of a given forward Libor rate. More detail on this can be found in Chap. 3 of Miglietta (2015).

Remark 3.4 The same drift condition as (3.33) is obtained in Theorem 3.3.3 of Miglietta (2015) (see also Theorem 3.4.3 therein). Miglietta (2015) derives this condition by using the fact that $\frac{p(t,T+\Delta)\,\Delta L(t;T,T+\Delta)}{B_t}$ is a martingale under the standard martingale measure Q with numéraire B_t. In fact, $p(t,T+\Delta)\,\Delta L(t;T,T+\Delta)$ is the price process of a traded asset, namely the time-t price of the floating leg in an FRA on the Libor that is setting at T and paying at $T+\Delta$. The fact that the two approaches lead to the same result is not surprising since, due to $\frac{dQ^{T+\Delta}}{dQ}\big|_{\mathscr{F}_t} = \frac{p(t,T+\Delta)}{B_t\,p(0,T+\Delta)}$, the Q-martingality of $\frac{p(t,T+\Delta)\,\Delta L(t;T,T+\Delta)}{B_t}$ is equivalent to the $Q^{T+\Delta}$-martingality of $\Delta L(t;T,T+\Delta)$.

3.2.2.2 Markovianity and Induced Short Rates (Vasiček-Type Volatility Structure)

The underlying quantities in the present Sect. 3.2.2 are the forward rates $f(t,T)$ and the spreads $g(t,T)$. Their dynamics are determined (see (3.30)) by their respective drifts $a(t,T), a^*(t,T)$ and volatilities $\sigma(t,T), \sigma^*(t,T)$. For the coefficients $a(t,T), \sigma(t,T)$ of the forward rates we have the explicit condition (3.12), while for $a^*(t,T), \sigma^*(t,T)$ we have a no-arbitrage condition implicitly contained in (3.33). Note that this condition does not uniquely determine $a^*(t,T)$ once $\sigma^*(t,T)$ and $\sigma(t,T)$ have been chosen. We shall see below that we are nevertheless able to obtain Markovianity and an induced model for the short-rate spread even without imposing the no-arbitrage condition (3.33).

From (3.30) and using the drift condition (3.12) for $a(t,T)$ in terms of $\sigma(t,T)$, but leaving $a^*(t,T)$ generic, we obtain

$$
f(t,T) = f(0,T) + \int_0^t \langle \Sigma(s,T), \sigma(s,T)\rangle ds + \int_0^t \sigma(s,T)dw_s \qquad (3.35)
$$

as well as

$$g(t, T) = g(0, T) + \int_0^t a^*(s, T)ds + \int_0^t \sigma^*(s, T)dw_s \qquad (3.36)$$

Similarly to Sect. 3.2.1.2, the representations in (3.35) and (3.36) are infinite-dimensional and again we shall investigate whether with a suitable choice of the volatilities $\sigma(t, T)$ and $\sigma^*(t, T)$ we may obtain finitely parametrized representations and Markovian models for the induced short-rate $r_t = f(t, t)$ and short rate spread $s_t = g(t, t)$. Again we limit ourselves for illustration purposes to just one example of Vasiček-type volatilities.

By analogy to the example in Sect. 3.2.1.2 also here we consider a two-dimensional Wiener process $w_t = (w_t^1, w_t^2)$ and, for given positive constants σ, σ^*, as well as non-zero $b^*, b \in \mathbb{R}$, assume the following volatility structures

$$\begin{aligned} \sigma(t, T) &= \left(\sigma e^{-b(T-t)}, \; 0\right) \\ \sigma^*(t, T) &= \left(0, \; \sigma^* e^{-b^*(T-t)}\right) \end{aligned} \qquad (3.37)$$

so that $f(t, T)$ is driven only by w_t^1 and $g(t, T)$ only by w_t^2. It also follows that

$$\begin{aligned} \Sigma(t, T) &= \int_t^T \sigma(t, u)du = \left(\frac{\sigma}{b}\left(1 - e^{-b(T-t)}\right), \; 0\right) \\ \Sigma^*(t, T) &= \int_t^T \sigma^*(t, u)du = \left(0, \; \frac{\sigma}{b^*}\left(1 - e^{-b^*(T-t)}\right)\right) \end{aligned} \qquad (3.38)$$

As in Sect. 3.2.1.2, here too we consider first the forward rates $f(t, T)$ and the induced short rate $r_t = f(t, t)$. Since $\Sigma(t, T)$ has here the same structure as the corresponding volatility in Sect. 3.2.1.2 (see (3.20)), we obtain the same results also here for $f(t, T)$ and r_t, namely we have again a Hull–White extended Vasiček model for r_t given by (3.25) thus leading to an affine Markovian term structure for the OIS bonds.

Consider next the induced model for the short-rate spread $s_t = g(t, t)$ without imposing a drift condition on $a^*(t, T)$. From (3.36) and (3.37) it follows that

$$g(t, T) = g(0, T) + \int_0^t a^*(s, T)ds + \sigma^* e^{-b^* T} \int_0^t e^{b^* s} dw_s^2 \qquad (3.39)$$

From here we obtain for the short-rate spread $s_t = g(t, t)$

$$\begin{aligned} s_t &= g(0, t) + \int_0^t a^*(s, t)ds + \sigma^* e^{-b^* t} \int_0^t e^{b^* s} dw_s^2 \\ &=: \mu(t) + \sigma^* e^{-b^* t} \int_0^t e^{b^* s} dw_s^2 \end{aligned} \qquad (3.40)$$

It implies

$$ds_t = \left(\mu'(t) - b^* \sigma^* e^{-b^* t} \int_0^t e^{b^* s} dw_s^2\right) dt + \sigma^* dw_t^2 \qquad (3.41)$$

$$= \left(\mu'(t) - b^*(s_t - \mu(t))\right) dt + \sigma^* dw_t^2$$

Defining then (under the assumption that $a^*(s, t)$ is differentiable in t)

$$
\begin{aligned}
\rho^*(t) &:= \mu(t) + \tfrac{1}{b^*}\,\mu'(t) \\
&= g(0, t) + \int_0^t a^*(s, t)\,ds \\
&\quad + \tfrac{1}{b^*}\left[\tfrac{\partial}{\partial t} g(0, t) + a^*(t, t) + \int_0^t \tfrac{\partial}{\partial t} a^*(s, t)\,ds \right]
\end{aligned}
\tag{3.42}
$$

we obtain that also the induced short-rate spread satisfies a Hull–White extended Vasiček model, namely

$$
ds_t = b^*(\rho^*(t) - s_t)dt + \sigma^*\, dw_t^2
\tag{3.43}
$$

and this has been obtained without previously imposing a no-arbitrage drift condition as we had done it for r_t. Note that the short rate r is driven solely by the Wiener process w^1 and the short-rate spread s solely by the Wiener process w^2, which is implied by the specific assumptions on the volatility structures and does not require w^1 and w^2 to be independent.

Remark 3.5 Other volatility structures may lead to analogous results, among them CIR models with stochastic volatility (see Crépey et al. 2012). In all these cases, both the induced models for the short rate and the short-rate spread are Markov and affine and so have all the benefits that are typical for affine term structure models when determining the prices of various interest rate derivatives (see Sects. 2.4 and 2.5).

3.3 Foreign Exchange and Credit Risk Analogy

In this section we shall elaborate on the approach mentioned in (**ii**) of the introduction to this chapter, namely the modeling of the dynamics of the actual Libor rates via the dynamics of fictitious bond prices that were generically denoted by $\bar{p}(t, T)$. Recall that the forward Libor rates are then given by the third equality in (3.5), namely $L(t, T, T + \Delta) = \tfrac{1}{\Delta}\left(\bar{\nu}_{t,T} - 1\right)$, with $\bar{\nu}_{t,T}$ given by (3.4).

3.3.1 Models and No-Arbitrage Conditions

In Chap. 2 we had followed the classical short-rate approach to obtain arbitrage-free prices $\bar{p}(t, T)$ by starting from the OIS short rate and adding to it a spread, analogously to what is done in credit risk, where the spread corresponds to the default intensity. Since here we are already at the higher level of the bond prices or, equivalently, the instantaneous forward rates, we have to derive conditions directly on the dynamics of $\bar{p}(t, T)$ or, equivalently, on $\bar{f}(t, T)$. To this effect, we describe below two possible specific interpretations of these fictitious bonds recalling that we had postulated (3.1). As mentioned in the introduction to this chapter, these

interpretations lead to pseudo no-arbitrage conditions since they are not implied by
absence of arbitrage in reference to the traded FRA contracts, which is ensured here
directly by model construction, due to the martingale property of the process $\bar{v}_{t,T}$
resulting from (3.4). This is why we shall refer to these conditions as pseudo no-
arbitrage conditions—in contrast to the previous section where the drift conditions
(3.16) and (3.33) were needed to ensure the martingale property of the processes
$G(\cdot, T, T + \Delta)$ and $\frac{p^{\Delta}(\cdot, T)}{p^{\Delta}(\cdot, T+\Delta)}$ in order to guarantee absence of arbitrage in reference
to the FRA contracts.

(a) *Interpretation of the $\bar{p}(\cdot, T)$-bonds as pre-default values of credit risky bonds.*

   This interpretation results from viewing the fictitious bonds $\bar{p}(t, T)$ as issued by
   an average Libor bank (see also Ametrano and Bianchetti 2013, Morini 2009,
   and Filipović and Trolle 2013) defaulting at a random time τ^*. In this sense
   in Crépey et al. (2012) no-arbitrage conditions are introduced using a credit-
   risk analogy. This interpretation was also already implicit in the previous use of
   $\bar{p}(\cdot, T)$ above and corresponds in some sense also to the interpretation given in
   Chapter 2, where the short-rate spread was introduced by analogy to the default
   intensity in credit risk. For this reason, we shall keep the symbol $\bar{p}(t, T)$ for this
   first interpretation.
(b) *Interpretation based on a foreign exchange analogy.*

   In this case we shall use the symbol $p^f(t, T)$ and interpret these bonds as denominated in a foreign currency. The relationships are the same as in (3.1) by
   replacing $\bar{p}(t, T)$ with $p^f(t, T)$. This approach has been suggested firstly by
   Bianchetti (2010) in the Libor market model setup (see also Miglietta 2015) and
   here we shall develop the same analogy for the HJM setup.

Both models for $\bar{p}(\cdot, T)$ and for $p^f(\cdot, T)$ can be obtained by proceeding as in
Sect. 3.1, where we used the "bar-notation". By analogy to the classical HJM approach, we shall again start by modeling directly the forward rates $\bar{f}(t, T) :=$
$-\frac{\partial}{\partial T} \log \bar{p}(t, T)$ via the spread $g(t, T)$. This yields the following model for the
OIS bond prices $p(t, T)$ and the $\bar{p}(t, T)$-bonds:

$$dp(t, T) = p(t, T) \left[(r_t - A(t, T) + \tfrac{1}{2}|\Sigma(t, T)|^2)dt - \Sigma(t, T)dw_t \right]$$

$$d\bar{p}(t, T) = \bar{p}(t, T) \left[(\bar{r}_t - \bar{A}(t, T) + \tfrac{1}{2}|\bar{\Sigma}(t, T)|^2)dt - \bar{\Sigma}(t, T)dw_t \right]$$

(3.44)

with $A(t, T)$ satisfying the drift condition (3.12).
   For the $p^f(\cdot, T)$-bonds we proceed in perfect analogy simply by replacing the
quantities denoted by a "bar" with respective quantities with superscript "f". It
means that, e.g., the volatility of the $p^f(\cdot, T)$-bond will be denoted $\Sigma^f(\cdot, T)$ and so
forth.
   We shall continue this section with the no-arbitrage drift conditions, separately for
each of the variants $\bar{p}(\cdot, T)$ and $p^f(\cdot, T)$. Note that these conditions are implied by

different interpretations of the bonds and are not directly implied by the no-arbitrage condition for the FRA contracts, as it was the case in Sect. 3.2.

3.3.1.1 No-Arbitrage Condition for the $\bar{p}(t, T)$-Bonds.

In order to deal with the bonds $\bar{p}(t, T)$, we view them as pre-default values of credit risky bonds $\bar{p}^R(t, T)$ with recovery $R \in [0, 1]$, following the approach suggested in Crépey et al. (2012). To arrive at a particularly convenient form of the no-arbitrage condition on $\bar{p}(t, T)$, Crépey et al. (2012) assume in fact the fractional recovery of a market value scheme for the risky bonds, which specifies that in case of default of the bond issuer, the fraction of the pre-default value of the bond is paid at default time τ^*. We obtain the following drift condition for the coefficients in the dynamics of $\bar{p}(t, T)$:

$$\bar{A}(t, T) = \frac{1}{2} |\bar{\Sigma}(t, T)|^2 \tag{3.45}$$

The derivation of this no-arbitrage condition is not straightforward without entering into a detailed analysis of the credit risk setup which is used in Crépey et al. (2012) and is thus omitted in order not to overburden the treatment. Very concisely, the value at maturity T of the risky bond in case of the fractional recovery of a market value scheme is given by

$$\bar{p}^R(T, T) = \mathbf{1}_{\{\tau^* > T\}} + \mathbf{1}_{\{\tau^* \leq T\}} R \, \bar{p}(\tau^* -, T) p(\tau^*, T)^{-1}$$

since receiving the amount $\mathbf{1}_{\{\tau^* \leq T\}} R \, \bar{p}(\tau^* -, T)$ at time τ^* is equivalent to receiving $\mathbf{1}_{\{\tau^* \leq T\}} R \, \bar{p}(\tau^* -, T) p(\tau^*, T)^{-1}$ at time T. Hence, the time-t price of the $\bar{p}^R(t, T)$ bond can be shown to be expressed as

$$\bar{p}^R(t, T) = \mathbf{1}_{\{\tau^* > t\}} \bar{p}(t, T) + \mathbf{1}_{\{\tau^* \leq t\}} R \, \bar{p}(\tau^*, T) p(\tau^*, T)^{-1} p(t, T)$$

under some standard assumptions from the intensity-based credit risk modeling. The above drift condition is then obtained by using some classical techniques from credit risk theory, where the details can be found in Sect. 2.3.2 of Crépey et al. (2012) for the case of a Lévy driving process, which in turn relies on Sect. 13.1.9 in Bielecki and Rutkowski (2001).

3.3.1.2 No-Arbitrage Condition for the $p^f(t, T)$-Bonds

Here we derive no-arbitrage conditions on the basis of a foreign exchange analogy developed in the HJM setup.

We start from the forward Libor rate, namely (see (3.1))

$$L(t; T, T + \Delta) = E^{T+\Delta} \{L(T; T, T + \Delta) \mid \mathscr{F}_t\}$$

but instead of $L(T; T, T + \Delta)$ expressed as in (3.1) in terms of $\bar{p}(t, T)$, we replace here $\bar{p}(t, T)$ with $p^f(t, T)$, namely

$$L(T; T, T + \Delta) = \frac{1}{\Delta} \left(\frac{1}{p^f(T, T + \Delta)} - 1 \right) \tag{3.46}$$

where the $p^f(t, T)$ are now interpreted as bonds denominated in a different, foreign currency. They can then be considered as traded assets in the foreign economy and this requires to impose no-arbitrage conditions on them.

In order to derive conditions for absence of arbitrage in the domestic and the foreign economies, we have to link the two markets. To this effect let a wealth X^f_t in the foreign currency be denoted by $\bar{X}^f_t$ when expressed in the domestic currency. It follows that $\bar{X}^f_t = S_t X^f_t$ where S_t is the spot exchange rate. In particular, $S_t p^f(t, T)$ is the price of a foreign bond in the domestic currency and can be considered as tradable asset in the domestic economy, which means that the discounted price process

$$\left(\frac{S_t p^f(t, T)}{B_t} \right)_{t \leq T}$$

must be a Q-martingale. This will give us the required no-arbitrage conditions on the bonds $p^f(t, T)$. Let us now define a model for the bonds $p^f(t, T)$. Since the $p^f(t, T)$ now play the role that was previously played by $\bar{p}(t, T)$, we assume for $p^f(t, T)$ a model completely analogous to that for $\bar{p}(t, T)$, having coefficients $A^f(t, T)$ and $\Sigma^f(t, T)$ in the dynamics of $p^f(t, T)$ under Q as given in (3.44). Having now the dynamics for $p(t, T)$ and $p^f(t, T)$, we have to postulate a dynamics for the spot exchange rate S_t, always under Q and so we make

Assumption 3.1 The spot exchange rate S_t satisfies

$$dS_t = S_t(\alpha_t dt + \beta_t dw_t) \tag{3.47}$$

with the drift $\alpha = r - r^f$, where r^f is the foreign short rate, and the volatility β an adapted processes satisfying the implicit integrability conditions.

The specific form of the drift above is needed to ensure the absence of arbitrage between the two markets, cf. for example Musiela and Rutkowski (2005, Sect. 14.1).

We can now state and prove

Proposition 3.3 *For the model (3.9), where $\bar{a}(t, T), \bar{\sigma}(t, T)$ and, correspondingly $\bar{A}(t, T), \bar{\Sigma}(t, T)$ are replaced by $a^f(t, T), \sigma^f(t, T), A^f(t, T), \Sigma^f(t, T)$ respectively, we have the no-arbitrage drift condition*

$$A^f(t, T) = \tfrac{1}{2} |\Sigma^f(t, T)|^2 - \langle \beta_t, \Sigma^f(t, T) \rangle \tag{3.48}$$

Proof Writing down the dynamics of $\frac{S_t p^f(t,T)}{B_t}$ with $p^f(t, T)$ as given in (3.44) and simplifying some terms in the drift, it follows

$$d\left(\frac{S_t p^f(t, T)}{B_t}\right) = \frac{S_t p^f(t, T)}{B_t}$$

$$\cdot \left(\left(\alpha_t + r_t^f - A^f(t, T) - r_t + \frac{1}{2}|\Sigma^f(t, T)|^2\right.\right.$$

$$\left.\left. - \langle\beta_t, \Sigma^f(t, T)\rangle\right)dt + (\beta_t - \Sigma^f(t, T))dw_t\right) \qquad (3.49)$$

Condition (3.48) then follows by setting the drift in (3.49) equal to zero and using Assumption 3.1. □

Remark 3.6 Note that condition (3.48) on the drift term $A^f(t, T)$ contains an additional term in comparison to the standard HJM drift condition, which depends on the volatility β_t of the exchange rate S_t and which is due to the transformation of the foreign bond price (i.e. fictitious bond price in the multiple curve setup) to the domestic economy (i.e. OIS-based economy in the multiple curve setup) via S_t.

Moreover, the condition from Assumption 3.1 imposes in addition a special form of the drift for the exchange rate, which has to be linked to the short domestic and foreign rates r and r^f (i.e. the short OIS and the short fictitious Libor rate).

Remark 3.7 Introducing now a foreign bank account

$$B_t^f = \exp\left(\int_0^t r_s^f ds\right)$$

related to the short rate r^f, one can easily verify that the discounted foreign bond price process $\frac{p^f(t,T)}{B_t^f}$ is a martingale with respect to a foreign martingale measure Q^f with numéraire B^f, which is linked to the measure Q via the following Radon-Nikodym density

$$\frac{dQ^f}{dQ}\bigg|_{\mathscr{F}_t} = \frac{S_t B_t^f}{B_t} = \mathcal{E}\left(\int_0^{\cdot} \beta_s dw_s\right)_t \qquad (3.50)$$

Remark 3.8 The forward exchange rate fixed at time t for date $T \geq t$ and denoted by $S(t, T)$ can be defined as the T-forward domestic price of a unit of foreign currency, given by

$$S(t, T) = \frac{S_t p^f(t, T)}{p(t, T)} \qquad (3.51)$$

Note that the spot exchange rate is then expressed as $S_t = S(t, t)$. The process $S(t, T)$ is a Q^T-martingale and it represents the price of a traded asset $S_t p^f(t, T)$ in the domestic economy using the $p(t, T)$-bond as numéraire. We can then introduce the foreign forward measure $Q^{f, T+\Delta}$ by its Radon-Nikodym density

$$\frac{dQ^{f, T+\Delta}}{dQ^{T+\Delta}}\bigg|_{\mathscr{F}_t} = \frac{S(t, T+\Delta)}{S(0, T+\Delta)} = const \frac{p^f(t, T+\Delta)}{p(t, T+\Delta)} S_t \qquad (3.52)$$

By straightforward calculations one finds that the foreign forward price process

$$\Pi^f(t, T) := \frac{p^f(t, T)}{p^f(t, T + \Delta)}$$

is a $Q^{f, T + \Delta}$-martingale as expected. We mention that in the LMM foreign exchange analogy presented in Bianchetti (2010) the forward exchange rate $S(t, T)$ is modeled directly under the forward measure Q^T.

3.4 Pricing of Interest Rate Derivatives

The interest rate derivatives that we are considering here are, as in the previous chapters, essentially derivatives whose underlying is the Libor rate. Based on Crépey et al. (2012) and Crépey et al. (2015a), we shall consider the pricing of such derivatives, limiting ourselves to clean prices that correspond to collateralized transactions with funding at the OIS rate (see the discussion in Chap. 1, in particular Sect. 1.2.3). We shall distinguish between linear derivatives, namely interest rate swaps and their variants, and nonlinear (optional) derivatives, in particular caps/floors and swaptions. For the linear derivatives we shall determine, in addition to their clean prices, also various associated spreads, in particular Libor-OIS swap spreads and basis swap spreads (see the definitions in Sects. 1.4.4 and 1.4.5). We shall see that the relationship (3.6) will allow us to set up the pricing of linear derivatives on a common basis for both modeling approaches described in Sects. 3.2 and 3.3 respectively. The specific calculations will however differ for each specific model and we shall show the resulting values next to one another. More precisely, we shall present the results for the HJM-LMM hybrid model for the process $G(t, T, T + \Delta)$ described in Sect. 3.2.1, for the HJM-LMM $p^\Delta(t, T)$-bond price model described in Sect. 3.2.2 and for the two models described in Sect. 3.3. The treatment of the last two models is, from the mathematical point of view, completely unified, hence, we shall in the sequel always treat these two models as one model type, i.e. the $\bar{p}(t, T)$-model. In order to obtain the corresponding results for the $p^f(t, T)$-model, all quantities denoted by a "bar" in the $\bar{p}(t, T)$-model have to be replaced with respective quantities with superscript f. The only difference between the two models is in the drift conditions, (3.45) for the $\bar{p}(t, T)$-model and (3.48) for the $p^f(t, T)$-model. Hence, in the remainder of the chapter we distinguish between three model types: the $G(t, T, T + \Delta)$-model from Sect. 3.2.1, the $p^\Delta(t, T)$-model from Sect. 3.2.2 and the $\bar{p}(t, T)$-model type applying to both models in Sect. 3.3.

Before coming to the pricing of the individual derivatives, let us start with some preliminaries related to the $\bar{p}(t, T)$-model type that will be useful below. In particular, we shall derive an explicit expression for the quantity

$$\bar{\nu}_{t, T} = E^{T + \Delta} \left\{ \frac{1}{\bar{p}(T, T + \Delta)} \middle| \mathscr{F}_t \right\}$$

which is the crucial quantity for derivative pricing in this model type, cf. (3.4). We first derive a more convenient expression for the bond prices as defined in (3.10) and (3.11). A direct integration leads to

$$p(t, T) = p(0, T) \exp\left[\int_0^t (r_s - A(s, T))ds - \int_0^t \Sigma(s, T) dw_s\right] \quad (3.53)$$

and

$$\bar{p}(t, T) = \bar{p}(0, T) \exp\left[\int_0^t (\bar{r}_s - \bar{A}(s, T))ds - \int_0^t \bar{\Sigma}(s, T) dw_s\right] \quad (3.54)$$

Setting $T = t$ in these two relations one obtains

$$B_t := \exp\left[\int_0^t r_s ds\right] = \frac{1}{p(0, t)} \exp\left[\int_0^t A(s, t)ds + \int_0^t \Sigma(s, t) dw_s\right] \quad (3.55)$$

with an analogous expression for $\bar{B}_t := \exp\left[\int_0^t \bar{r}_s ds\right]$. Substituting the latter expressions back into (3.53) and (3.54) one obtains

$$\begin{aligned}
p(t, T) &= \frac{p(0,T)}{p(0,t)} \exp\left[\int_0^t (A(s, t) - A(s, T))ds + \int_0^t (\Sigma(s, t) - \Sigma(s, T))dw_s\right] \\
&= \frac{p(0,T)}{p(0,t)} \exp\left[\frac{1}{2}\int_0^t \left(|\Sigma(s, t)|^2 - |\Sigma(s, T)|^2\right) ds\right] + \int_0^t (\Sigma(s, t) - \Sigma(s, T))dw_s
\end{aligned} \quad (3.56)$$

where the second equality follows from the no-arbitrage drift condition (3.12), and

$$\bar{p}(t, T) = \frac{\bar{p}(0, T)}{\bar{p}(0, t)} \exp\left[\int_0^t (\bar{A}(s, t) - \bar{A}(s, T))ds + \int_0^t (\bar{\Sigma}(s, t) - \bar{\Sigma}(s, T))dw_s\right] \quad (3.57)$$

The expressions in (3.56) and (3.57) are under the martingale measure Q with the Wiener process w_t. Below we shall also need these expressions under various forward measures, in particular under $Q^{T+\Delta}$ with the Wiener process $w_t^{T+\Delta}$ that is obtained from w_t by the translation specified in (3.17).

For their use in the sequel, we can now derive more explicit expressions for $\nu_{t,T}$ and $\bar{\nu}_{t,T}$ that were first introduced in (2.56) and (2.50) and where $\bar{\nu}_{t,T}$ is also recalled in (3.4). We do it here explicitly for the model expressed by $A(t, T)$, $\bar{A}(t, T)$, $\Sigma(t, T)$ and $\bar{\Sigma}(t, T)$. For the alternative model expressed by $A^f(t, T)$ and $\Sigma^f(t, T)$, the results are completely analogous.

For $\nu_{t,T}$ the expression under Q is given by

$$\begin{aligned}
\nu_{t,T} &= \frac{p(t,T)}{p(t,T+\Delta)} = \frac{p(0,T)}{p(0,T+\Delta)} \\
&\cdot \exp\left[\frac{1}{2}\int_0^t \left(|\Sigma(s, T + \Delta)|^2 - |\Sigma(s, T)|^2\right) ds + \int_0^t (\Sigma(s, T + \Delta) \right. \\
&\left. - \Sigma(s, T))dw_s\right]
\end{aligned} \quad (3.58)$$

and, under $Q^{T+\Delta}$, one obtains it immediately by applying (3.17).

For what concerns $\bar{\nu}_{t,T}$, let us for convenience of notation introduce a process Ψ^T satisfying, under Q,

$$d\Psi_t^T = (\bar{\Sigma}(t, T + \Delta) - \bar{\Sigma}(t, T))dw_t \tag{3.59}$$

From the definition of $\bar{\nu}_{t,T}$ and (3.57), as well as (3.17), we now have

$$
\begin{aligned}
\bar{\nu}_{t,T} &= E^{T+\Delta}\left\{ \frac{1}{\bar{p}(T,T+\Delta)} \mid \mathscr{F}_t \right\} \\
&= \frac{\bar{p}(0,T)}{\bar{p}(0,T+\Delta)} \exp\left[\int_0^T \left(\bar{A}(s, T + \Delta) - \bar{A}(s, T) \right) ds \right] \\
&\quad \cdot \exp\left[\int_0^t \left(\bar{\Sigma}(s, T + \Delta) - \bar{\Sigma}(s, T) \right) dw_s \right] \\
&\quad \cdot E^{T+\Delta}\left\{ \exp\left[\int_t^T \left(\bar{\Sigma}(s, T + \Delta) - \bar{\Sigma}(s, T) \right) dw_s \right] \mid \mathscr{F}_t \right\} \\
&= c_T e^{\Psi_t^T} \exp\left[-\int_t^T \langle \Sigma(s, T + \Delta), \bar{\Sigma}(s, T + \Delta) - \bar{\Sigma}(s, T) \rangle ds \right] \\
&\quad \cdot E^{T+\Delta}\left\{ \exp\left[\int_t^T \left(\bar{\Sigma}(s, T + \Delta) - \bar{\Sigma}(s, T) \right) dw_s^{T+\Delta} \right] \mid \mathscr{F}_t \right\} \\
&= c_T e^{\Psi_t^T} \exp\left[-\int_t^T \langle \Sigma(s, T + \Delta), \bar{\Sigma}(s, T + \Delta) - \bar{\Sigma}(s, T) \rangle ds \right] \\
&\quad \cdot \exp\left[\tfrac{1}{2} \int_t^T |\bar{\Sigma}(s, T + \Delta) - \bar{\Sigma}(s, T)|^2 ds \right] \\
&= c_T e^{\Psi_t^T} \exp\left[\tfrac{1}{2} \int_t^T \left(|\bar{\Sigma}(s, T + \Delta) - \bar{\Sigma}(s, T) - \Sigma(s, T + \Delta)|^2 - |\Sigma(s, T + \Delta)|^2 \right) ds \right]
\end{aligned}
\tag{3.60}
$$

where we have used the shorthand notation

$$c_T := \frac{\bar{p}(0, T)}{\bar{p}(0, T + \Delta)} \exp\left[\int_0^T \left(\bar{A}(s, T + \Delta) - \bar{A}(s, T) \right) ds \right] \tag{3.61}$$

In this expression the factor $\frac{\bar{p}(0,T)}{\bar{p}(0,T+\Delta)}$ has to be considered as a parameter to be calibrated and $\bar{A}(t, T)$ has to satisfy the no-arbitrage drift conditions for the model under consideration ((3.45), respectively (3.48)).

In contrast to this, in the model for $p^\Delta(\cdot, T)$-bonds, we directly obtain (see (3.29)) by using (3.54) and replacing all the quantities therein with the corresponding ones with superscript Δ

$$
\begin{aligned}
\frac{p^\Delta(t, T)}{p^\Delta(t, T + \Delta)} &= \frac{p^\Delta(0, T)}{p^\Delta(0, T + \Delta)} \exp\left[\int_0^t \left(A^\Delta(s, T + \Delta) - A^\Delta(s, T) \right) ds \right] \\
&\quad \cdot \exp\left[\int_0^t \left(\Sigma^\Delta(s, T + \Delta) - \Sigma^\Delta(s, T) \right) dw_s \right] \\
&= c_T^\Delta \Psi_t^{T,\Delta}
\end{aligned}
\tag{3.62}
$$

with

$$c_T^\Delta := \frac{p^\Delta(0, T)}{p^\Delta(0, T + \Delta)} \exp\left[\int_0^t \left(A^\Delta(s, T + \Delta) - A^\Delta(s, T) \right) ds \right]$$

and (compare also (3.59))

$$d\Psi_t^{T,\Delta} = (\Sigma^\Delta(t, T + \Delta) - \Sigma^\Delta(t, T))dw_t$$

Here we emphasize that the initial ratio $\frac{p^\Delta(0,T)}{p^\Delta(0,T+\Delta)}$ is a market observable quantity, as opposed to $\frac{\bar{p}(0,T)}{\bar{p}(0,T+\Delta)}$ and $\frac{p^f(0,T)}{p^f(0,T+\Delta)}$ that have to be considered as parameters to be calibrated to the market.

3.4.1 Linear Derivatives: Interest Rate Swaps

From (1.27), the price of a Libor indexed payer swap can be expressed as follows (the various δ_k correspond to what for the single tenor case we had denoted by Δ)

$$P^{Sw}(t; T_0, T_n, R, N) = N \sum_{k=1}^n \delta_k p(t, T_k) E^{Q^k} \{L(T_{k-1}; T_{k-1}, T_k) - R|\mathcal{F}_t\}$$
$$= \begin{cases} N \sum_{k=1}^n p(t, T_k) (G(t; T_{k-1}, T_k) - \delta_k R) \\ N \sum_{k=1}^n p(t, T_k) \left(\frac{p^\Delta(t,T_{k-1})}{p^\Delta(t,T_k)} - \bar{R}_k\right) \\ N \sum_{k=1}^n p(t, T_k) (\bar{\nu}_{t,T_{k-1}} - \bar{R}_k) \end{cases}$$
$$(3.63)$$

with $\bar{R}_k := 1 + \delta_k R$, for the three model classes of Sects. 3.2.1, 3.2.2 and 3.3, respectively.

For the two models of Sects. 3.2.1 and 3.2.2 we can directly determine the various $G(t; T_{k-1}, T_k)$, respectively $\frac{p^\Delta(t,T_{k-1})}{p^\Delta(t,T_k)}$ (see e.g. the Vasiček example in Sects. 3.2.1.2 and 3.2.2.2) and thus also the price of the swap. For the models described in Sect. 3.3, we first determine $\bar{\nu}_{t,T_{k-1}} = E^{Q^{T_k}} \left\{ \frac{1}{\bar{p}(T_{k-1},T_k)} \mid \mathcal{F}_t \right\}$, just as it was the case in the short-rate approach of Chap. 2, and then follow the third alternative in (3.63) to determine the price of the swap (recall that in (3.60) we have derived an explicit expression for $\bar{\nu}_{t,T_{k-1}}$). For each model type we can, therefore, obtain the expressions for the price of the Libor indexed swap and the corresponding swap rate $R(t; T_0, T_n)$, namely the rate R that makes equal to zero the time-t value $P^{Sw}(t; T_0, T_n, R, N)$ as it results from (3.63). Corresponding to the three equivalent alternative expressions in (3.63), we obtain in fact

$$R(t; T_0, T_n) = \begin{cases} \frac{\sum_{k=1}^n p(t,T_k)G(t;T_{k-1},T_k)}{\sum_{k=1}^n \delta_k p(t,T_k)} \\ \frac{\sum_{k=1}^n p(t,T_k)\left(\frac{p^\Delta(t,T_{k-1})}{p^\Delta(t,T_k)}-1\right)}{\sum_{k=1}^n \delta_k p(t,T_k)} \\ \frac{\sum_{k=1}^n p(t,T_k)(\bar{\nu}_{t,T_{k-1}}-1)}{\sum_{k=1}^n \delta_k p(t,T_k)} \end{cases} \quad (3.64)$$

3.4.2   Linear Derivatives: Specific Swaps and Ensuing Spreads

While in the previous subsection we considered the standard interest rate swap in the multi-curve HJM models, here we consider some specific swaps, some of which, such as the basis swaps, were motivated by the crisis itself. We consider also the related spreads that before the crisis would have been negligible.

3.4.2.1   FRA Rates

We start, for completeness, from the simplest swap, namely the standard forward rate agreement (FRA) (see Sects. 1.4.1 and 2.3) that can be seen as a particular case of the standard interest rate swap for $n = 1$. Its price can thus be expressed as

$$P^{FRA}(t; T, T + \Delta, R, N) = \begin{cases} Np(t, T + \Delta)\,(G(t; T, T + \Delta) - \Delta\,R) \\ Np(t, T + \Delta)\left(\frac{p^{\Delta}(t,T)}{p^{\Delta}(t,T+\Delta)} - (1 + \Delta\,R)\right) \\ Np(t, T + \Delta)\left(\bar{\nu}_{t,T} - (1 + \Delta\,R)\right) \end{cases} \quad (3.65)$$

for the two model types in Sects. 3.2.1, 3.2.2 and that of Sect. 3.3 respectively. It follows that the FRA rate $R^{FRA}(t; T, T + \Delta)$, namely the rate R that makes the time-t value $P^{FRA}(t; T, T + \Delta, R, N)$ equal to zero, is given by

$$R^{FRA}(t; T, T + \Delta) = \begin{cases} \frac{1}{\Delta}\,G(t; T, T + \Delta) \\ \frac{1}{\Delta}\left(\frac{p^{\Delta}(t,T)}{p^{\Delta}(t,T+\Delta)} - 1\right) \\ \frac{1}{\Delta}\left(\bar{\nu}_{t,T} - 1\right) \end{cases} \quad (3.66)$$

Note that $R^{FRA}(t; T, T + \Delta) = L(t; T, T + \Delta)$, cf. Definition 1.2. Recalling that the forward OIS rate as defined in (1.16) is given by

$$F(t; T, T + \Delta) = \frac{1}{\Delta}\left(\frac{p(t, T)}{p(t, T + \Delta)} - 1\right), \quad t \le T \quad (3.67)$$

we have that the spread of the FRA rate R^{FRA} over the forward OIS rate, which was defined in (1.35) and referred to as the forward Libor-OIS spread, is

$$S(t; T, T + \Delta) = \begin{cases} \frac{1}{\Delta}\left(G(t; T, T + \Delta) + 1 - \frac{p(t,T)}{p(t,T+\Delta)}\right) \\ \frac{1}{\Delta}\left(\frac{p^{\Delta}(t,T)}{p^{\Delta}(t,T+\Delta)} - \frac{p(t,T)}{p(t,T+\Delta)}\right) \\ \frac{1}{\Delta}\left(\bar{\nu}_{t,T} - \frac{p(t,T)}{p(t,T+\Delta)}\right) \end{cases} \quad (3.68)$$

3.4.2.2 OIS Swaps and Spreads

These swaps have been discussed in Sect. 1.4.4, from where it also follows that the OIS rate, namely the rate R such that the value of the OIS at time t is equal to zero, is given by (see (1.32))

$$R^{OIS}(t; T_0, T_n) = \frac{p(t, T_0) - p(t, T_n)}{\sum_{k=1}^{n} \delta_k p(t, T_k)} \tag{3.69}$$

We can now obtain the Libor-OIS swap spread at time t (see (1.37)) as

$$R(t; T_0, T_n) - R^{OIS}(t; T_0, T_n) = \begin{cases} \frac{\sum_{k=1}^{n} p(t,T_k)G(t;T_{k-1},T_k) - p(t,T_0) + p(t,T_n)}{\sum_{k=1}^{n} \delta_k p(t,T_k)} \\[3mm] \frac{\sum_{k=1}^{n} p(t,T_k)\left(\frac{p^{\Delta}(t,T_{k-1})}{p^{\Delta}(t,T_k)} - 1\right) - p(t,T_0) + p(t,T_n)}{\sum_{k=1}^{n} \delta_k p(t,T_k)} \\[3mm] \frac{\sum_{k=1}^{n} p(t,T_k)\left(\bar{\nu}_{t,T_{k-1}} - 1\right) - p(t,T_0) + p(t,T_n)}{\sum_{k=1}^{n} \delta_k p(t,T_k)} \end{cases} \tag{3.70}$$

for the three model types respectively.

3.4.2.3 Basis Swaps and Spreads

These swaps were discussed in Sect. 1.4.5. Calculating the swap spread as defined in (1.40) in the three specific model types, one obtains

$$S^{BSw}(t; \mathcal{T}^1, \mathcal{T}^2) = \begin{cases} \frac{\sum_{i=1}^{n_1} p(t,T_i^1)G(t;T_{i-1}^1,T_i^1) - \sum_{j=1}^{n_2} p(t,T_j^2)G(t;T_{j-1}^2,T_j^2)}{\sum_{j=1}^{n_2} \delta_j^2 p(t,T_j^2)} \\[3mm] \frac{\sum_{i=1}^{n_1} p(t,T_i^1)\left(\frac{p^{\Delta}(t,T_{i-1}^1)}{p^{\Delta}(t,T_i^1)} - 1\right) - \sum_{j=1}^{n_2} p(t,T_j^2)\left(\frac{p^{\Delta}(t,T_{j-1}^2)}{p^{\Delta}(t,T_j^2)} - 1\right)}{\sum_{j=1}^{n_2} \delta_j^2 p(t,T_j^2)} \\[3mm] \frac{\sum_{i=1}^{n_1} p(t,T_i^1)\left(\bar{\nu}_{t,T_{i-1}^1} - 1\right) - \sum_{j=1}^{n_2} p(t,T_j^2)\left(\bar{\nu}_{t,T_{j-1}^2} - 1\right)}{\sum_{j=1}^{n_2} \delta_j^2 p(t,T_j^2)} \end{cases} \tag{3.71}$$

3.4.3 Caps and Floors

We shall concentrate here on the pricing of a caplet, from where the pricing of an entire cap, respectively floor, can be easily deduced. We derive the price of a generic caplet in parallel for the case of the models of Sects. 3.2.1 and 3.2.2 and the model type of Sect. 3.3, namely

$$P^{Cpl}(t; T + \Delta, K) = \Delta p(t, T + \Delta) E^{Q^{T+\Delta}} \left\{ (L(T; T, T + \Delta) - K)^+ \,|\mathscr{F}_t \right\}$$

$$= \begin{cases} p(t, T + \Delta) E^{Q^{T+\Delta}} \left\{ (G(T; T, T + \Delta) - \bar{K}^1)^+ \,|\mathscr{F}_t \right\} \\[2mm] p(t, T + \Delta) E^{Q^{T+\Delta}} \left\{ \left(\frac{1}{p^\Delta(T, T+\Delta)} - \bar{K}^2 \right)^+ \,|\mathscr{F}_t \right\} \\[2mm] p(t, T + \Delta) E^{Q^{T+\Delta}} \left\{ \left(\frac{1}{\bar{p}(T, T+\Delta)} - \bar{K}^2 \right)^+ \,|\mathscr{F}_t \right\} \end{cases}$$

$$(3.72)$$

where $\bar{K}^1 := \Delta K$, $\bar{K}^2 := 1 + \Delta K$.

Noting that in the above equation, the payoff of the caplet takes exactly the same form for the model of Sect. 3.2.2, as well as for the models in Sect. 3.3, we proceed below only with the treatment of the first two cases; the third one is then completely analogous to the second one.

In the three subsections below we shall discuss three possible approaches for the Wiener driven models, namely one leading to a Black-Scholes-type formula, one based on Fourier transform methods and finally one for the special structures of the volatility for which one can derive a corresponding short-rate model (see e.g. the Vasiček-type structures in Sects. 3.2.1.2 and 3.2.2.2); the approach in this latter case becomes then analogous to that in Chap. 2.

3.4.3.1   Black-Scholes-Type Approach

We shall now evaluate the right-hand side of (3.72) for which, in order to avoid needless formal complication, we shall set $t = 0$. Furthermore, for this subsection and the next one it is convenient to introduce two random variables X_1 and X_2, defined via

$$e^{X_1} := G(T; T, T + \Delta) \quad ; \quad e^{X_2} := \frac{1}{p^\Delta(T, T + \Delta)} \qquad (3.73)$$

so that, for $t = 0$, (3.72) becomes

$$P^{Cpl}(0; T + \Delta, K) = \begin{cases} p(0, T + \Delta) E^{Q^{T+\Delta}} \left\{ (e^{X_1} - \bar{K}^1)^+ \right\} \\[2mm] p(0, T + \Delta) E^{Q^{T+\Delta}} \left\{ (e^{X_2} - \bar{K}^2)^+ \right\} \end{cases} \qquad (3.74)$$

for the models of Sects. 3.2.1 and 3.2.2 respectively.

To evaluate the right-hand side of (3.74) we need the distributions of X_1 and X_2 under $Q^{T+\Delta}$ that, for our models, turn out to be Gaussian. The evaluation in (3.74) can thus be obtained via a Black-Scholes-type formula as shown in Proposition 3.4 below.

We start from X_1, where the underlying model is (3.15). Using the transformation of the Wiener process when passing from Q to $Q^{T+\Delta}$ (see (3.17)), we obtain the following expression under $Q^{T+\Delta}$

$$X_1 := \log G(T; T, T + \Delta)$$
$$= \log G(0; T, T + \Delta) - \tfrac{1}{2} \int_0^T |\sigma(s; T, T + \Delta)|^2 ds + \int_0^T \sigma(s; T, T + \Delta) dw_s^{T+\Delta} \quad (3.75)$$

where we have already taken into account the no-arbitrage condition expressed by the drift condition (3.16). Defining

$$\Gamma_T^1 := \int_0^T |\sigma(s; T, T + \Delta)|^2 ds$$
$$\gamma_T^1 := \log G(0; T, T + \Delta) - \tfrac{1}{2} \Gamma_T^1 \quad (3.76)$$

we may rewrite (3.75) as

$$X_1 = \gamma_T^1 + Y_T^1 \quad (3.77)$$

with $Y_T^1 \sim N(0, \Gamma_T^1)$, namely a Gaussian random variable with mean zero and co-variance Γ_T^1.

Coming to X_2, the underlying model is (3.32). Using again the transformation of the Wiener process when passing from Q to $Q^{T+\Delta}$ and re-writing (3.32) in a form analogous to the representation in (3.57), we obtain the following expression under $Q^{T+\Delta}$

$$X_2 := \log \tfrac{1}{p^\Delta(T, T+\Delta)}$$
$$= \log \tfrac{p^\Delta(0,T)}{p^\Delta(0,T+\Delta)} + \int_0^T \left(A^\Delta(s, T + \Delta) - A^\Delta(s, T) \right) ds$$
$$- \int_0^T \langle \Sigma(s, T + \Delta), \Sigma^\Delta(s, T + \Delta) - \Sigma^\Delta(s, T) \rangle ds \quad (3.78)$$
$$+ \int_0^T \left(\Sigma^\Delta(s, T + \Delta) - \Sigma^\Delta(s, T) \right) dw_s^{T+\Delta}$$

The coefficients $A^\Delta(s, T)$ have to satisfy the no-arbitrage drift condition (3.33). Recall that, when a similar expression is derived for the case of $\bar{p}(t, T)$-bonds, respectively $p^f(t, T)$-bonds, the drift conditions on the coefficients $\bar{A}(t, T)$ and $A^f(t, T)$ are expressed in (3.45) and (3.48) respectively.

Define, analogously to (3.76)

$$\Gamma_T^2 := \int_0^T |\Sigma^\Delta(s, T + \Delta) - \Sigma^\Delta(s, T)|^2 ds$$
$$\gamma_T^2 := \log \tfrac{p^\Delta(0,T)}{p^\Delta(0,T+\Delta)} + \int_0^T \left(A^\Delta(s, T + \Delta) - A^\Delta(s, T) \right) ds \quad (3.79)$$
$$- \int_0^T \langle \Sigma(s, T + \Delta), \Sigma^\Delta(s, T + \Delta) - \Sigma^\Delta(s, T) \rangle ds$$

we can rewrite (3.78) as

$$X_2 = \gamma_T^2 + Y_T^2 \quad (3.80)$$

with $Y_T^2 \sim N(0, \Gamma_T^2)$.

In this way we have the same expression $X_i = \gamma_T^i + Y_T^i$ $(i = 1, 2)$ for the two cases, which allows us to state the following Proposition in a unified form for all models. We have in fact the following analog of Proposition 2.4, namely

Proposition 3.4 *The price, at $t = 0$, of a caplet for the interval $[T, T + \Delta]$ with strike K on the Libor rate $L(T; T, T + \Delta)$ can be computed as*

$$P^{Cpl}(0; T + \Delta, K) = p(0, T + \Delta)e^{\gamma_T^i}\left[e^{\frac{\Gamma_T^i}{2}} N\left(\frac{1}{\sqrt{\Gamma_T^i}}\left(\log\left(\frac{1}{\bar{K}^i}\right) + \Gamma_T^i\right)\right)\right.$$
$$\left. - \tilde{K}^i N\left(\frac{1}{\sqrt{\Gamma_T^i}}\log\left(\frac{1}{\tilde{K}^i}\right)\right)\right] \qquad (3.81)$$

where, depending on the model, we set $i = 1$ or $i = 2$, $N(\cdot)$ is the cumulative standard Gaussian distribution function and $\tilde{K}^i := \bar{K}^i e^{-\gamma_T^i}$ with $\bar{K}^i$ as in (3.72).

Proof (analogous to the proof of Proposition 2.4). With the above notations we have

$$P^{Cpl}(0; T + \Delta, K)$$
$$= p(0, T + \Delta)E^{T+\Delta}\left\{\left(e^{X_i} - \bar{K}^i\right)^+\right\}$$
$$= p(0, T + \Delta)\, E^{T+\Delta}\left\{e^{\gamma_T^i}e^{Y_T^i}1_{\{e^{\gamma_T^i}e^{Y_T^i} > \bar{K}^i\}}\right\} - \bar{K}^i E^{T+\Delta}\left\{1_{\{e^{\gamma_T^i}e^{Y_T^i} > \bar{K}^i\}}\right\}$$
$$= p(0, T + \Delta)e^{\gamma_T^i}\left[E^{T+\Delta}\left\{e^{Y_T^i}1_{\{Y_T^i > \log(\tilde{K}^i)\}}\right\} - \tilde{K}^i E^{T+\Delta}\left\{1_{\{Y_T^i > \log(\tilde{K}^i)\}}\right\}\right]$$
$$= p(0, T + \Delta)e^{\gamma_T^i}$$
$$\cdot\left[\int_{\frac{\log(\tilde{K}^i)}{\sqrt{\Gamma_T^i}}}^{\infty} e^{x\sqrt{\Gamma_T^i}}\frac{1}{\sqrt{2\pi}}e^{-\frac{1}{2}x^2}dx - \tilde{K}^i P\left\{N(0, 1) > \frac{1}{\sqrt{\Gamma_T^i}}\log\left(\tilde{K}^i\right)\right\}\right]$$
$$= p(0, T + \Delta)e^{\gamma_T^i}$$
$$\cdot\left[e^{\frac{\Gamma_T^i}{2}}\frac{1}{\sqrt{2\pi}}\int_{\frac{\log(\tilde{K}^i)}{\sqrt{\Gamma_T^i}}}^{\infty} e^{-\frac{(x-\sqrt{\Gamma_T^i})^2}{2}}dx - \tilde{K}^i N\left(\frac{1}{\sqrt{\Gamma_T^i}}\log\left(\frac{1}{\tilde{K}^i}\right)\right)\right]$$
$$= p(0, T + \Delta)e^{\gamma_T^i}\left[e^{\frac{\Gamma_T^i}{2}} N\left(\frac{1}{\sqrt{\Gamma_T^i}}\left(\log\left(\frac{1}{\bar{K}^i}\right) + \Gamma_T^i\right)\right)\right.$$
$$\left. - \tilde{K}^i N\left(\frac{1}{\sqrt{\Gamma_T^i}}\log\left(\frac{1}{\tilde{K}^i}\right)\right)\right] \qquad (3.82)$$

$\square$

Remark 3.9 Note that the expression for the caplet price (3.81) in case of the model (3.15) (which corresponds to setting $i = 1$ in the formula) is almost exactly Black's formula for caplet prices in the log-normal Libor market model. This is obvious since in this case, the modeling object is the process $G(\cdot; T, T + \Delta)$, which is the forward Libor rate multiplied by the length of the interval Δ. Hence, the caplet corresponds to a call option on the underlying $G(T; T, T + \Delta)$, which, according to (3.75), has log-normal distribution under the forward measure $Q^{T+\Delta}$.

3.4.3.2 Fourier Transform-Based Approach

To evaluate the right-hand sides in (3.74), following Crépey et al. (2012) and Crépey et al. (2015a), one may also use the Fourier transform method (see Carr and Madan 1999 or Eberlein et al. 2010). To this effect notice that the generic payoff of the caplet, namely $g(x) := (e^x - \bar{K})^+$, has the generalized Fourier transform

$$\hat{g}(x) = \int_{\mathbb{R}} e^{izx} g(x)dx = \frac{\bar{K}^{1+iz}}{iz(1+iz)} \tag{3.83}$$

for $z \in \mathbb{C}$ with $Im(z) > 1$. We have now the following result that applies to both situations considered in (3.74) above, namely

Proposition 3.5 *The price, at $t = 0$, of a caplet for the interval $[T, T + \Delta]$ with strike K on the Libor rate $L(T; T, T + \Delta)$ can be computed as*

$$
\begin{aligned}
P^{Cpl}(0; T + \Delta, K) &= \frac{p(0,T+\Delta)}{2\pi} \int_{\mathbb{R}} \hat{g}(i\mathscr{R} - v) \, M_X^{T+\Delta}(\mathscr{R} + iv)dv \\
&= \frac{p(0,T+\Delta)}{2\pi} \int_{\mathbb{R}} \frac{\bar{K}^{1-iv-\mathscr{R}} M_X^{T+\Delta}(\mathscr{R}+iv)dv}{(\mathscr{R}+iv)(\mathscr{R}+iv-1)} \, dv
\end{aligned}
\tag{3.84}
$$

where $M_X^{T+\Delta}(z)$ is the moment generating function $M_X^{T+\Delta}(z) := E^{Q^{T+\Delta}} \{e^{zX}\}$ with $X = X_1$ or $X = X_2$ respectively and where $\mathscr{R} \geq 1$.

Remark 3.10 Although the right-hand side in (3.84) depends formally on $\mathscr{R}$, it is actually independent of the particular choice of a value for $\mathscr{R}$. On the other hand, this value affects however the complexity of a numerical evaluation of the integral.

It remains thus to determine $M_X^{T+\Delta}(z)$ for $X = X_1$ and $X = X_2$ respectively.
We start from the case X_1. We have from (3.75) and performing the measure transformation from $Q^{T+\Delta}$ to Q

$$
\begin{aligned}
E^{Q^{T+\Delta}} \{e^{zX_1}\} &= E^Q \{\mathscr{L}_T^{T+\Delta} e^{zX_1}\} \\
&= \exp\left[z\left(\log G(0; T, T + \Delta) + \int_0^T \alpha(s; T, T + \Delta)ds \right) \right] \\
&\quad \cdot E^Q \Big\{ \exp\Big[\int_0^T -\Sigma(s, T + \Delta)dw_s - \tfrac{1}{2} \int_0^T |\Sigma(s, T + \Delta)|^2 ds \\
&\qquad + z \int_0^T \sigma(s; T, T + \Delta)dw_s \Big] \Big\} \\
&= \exp\Big[z\Big(\log G(0; T, T + \Delta) \\
&\qquad + \int_0^T \Big(-\tfrac{1}{2}|\sigma(s; T, T + \Delta)|^2 + \langle \sigma(s; T, T + \Delta)\Sigma(s, T + \Delta)\rangle \Big) ds \Big) \\
&\qquad + \tfrac{1}{2} \int_0^T |z\sigma(s; T, T + \Delta) - \Sigma(s, T + \Delta)|^2 ds - \tfrac{1}{2} \int_0^T |\Sigma(s, T + \Delta)|^2 ds \Big] \\
&= \exp\Big[z\Big(\log G(0; T, T + \Delta) + \tfrac{1}{2}(z - 1) \int_0^T |\sigma(s; T, T + \Delta)|^2 ds \Big) \Big]
\end{aligned}
\tag{3.85}
$$

where the right-hand side can be computed explicitly.

Coming next to the case $X = X_2$, we have from the expression for X_2 in (3.78) and performing the same measure transformation as before from $Q^{T+\Delta}$ to Q

$$
\begin{aligned}
E^{Q^{T+\Delta}}\left\{e^{zX_2}\right\} &= E^Q\left\{\mathscr{L}_T^{T+\Delta}e^{zX_2}\right\} \\
&= \exp\left[z\left(\log\frac{p^\Delta(0,T)}{p^\Delta(0,T+\Delta)} + \int_0^T \left(A^\Delta(s,T+\Delta) - A^\Delta(s,T)\right)ds\right)\right] \\
&\quad \cdot E^Q\left\{\exp\left[\int_0^T -\Sigma(s,T+\Delta)dw_s - \frac{1}{2}\int_0^T |\Sigma(s,T+\Delta)|^2 ds\right.\right. \\
&\qquad \left.\left. +z\int_0^T \left(\Sigma^\Delta(s,T+\Delta) - \Sigma^\Delta(s,T)\right)dw_s\right]\right\} \\
&= \exp\left[z\left(\log\frac{p^\Delta(0,T)}{p^\Delta(0,T+\Delta)} + \int_0^T \left(A^\Delta(s,T+\Delta) - A^\Delta(s,T)\right)ds\right)\right. \\
&\quad +\frac{1}{2}\int_0^T |z\left(\Sigma^\Delta(s,T+\Delta) - \Sigma^\Delta(s,T)\right) - \Sigma(s,T+\Delta)|^2 ds \\
&\quad \left. -\frac{1}{2}\int_0^T |\Sigma(s,T+\Delta)|^2 ds\right]
\end{aligned}
$$

(3.86)

where the right-hand side can be computed explicitly. Further simplifications can be obtained by inserting the no-arbitrage condition (3.33).

3.4.3.3 Exponentially Affine Structure

Here we recall that, for the model of Sect. 3.2 with Vašiček-type volatilities, we had in Sect. 3.2.1.2 derived the following expression for the process $G(\cdot; T, T+\Delta)$ (see (3.26) and the shorthand notations used there)

$$G(t; T, T+\Delta) = \exp[m(t,T) + n(t,T)r_t + n^*(t,T)q_t] \qquad (3.87)$$

given in terms of the Markovian factors r_t and q_t which, under Q, satisfy (see (3.25) and (3.28))

$$
\begin{cases}
dr_t = b\left(\rho(t) - r_t\right)dt + \sigma\,dw_t^1 \\
dq_t = -b^* q_t dt + dw_t^2, \quad q_0 = 0
\end{cases}
\qquad (3.88)
$$

This allows for an alternative derivation of the price of a caplet as specified in the first equivalent representation in (3.72), which thus becomes analogous to the pricing of a call option on a bond for the case of an exponentially affine term structure. This is in line with the fact that, since for Vašiček-type volatilities one obtains a model for a corresponding short rate and an additional Markovian factor that may be seen as playing the role of a short-rate spread, we can proceed as in the affine short-rate models of Chap. 2.

3.4.4 Swaptions

Recall from Sect. 1.4.7 that the price, at $t \leq T$, of a payer swaption with maturity T, which we assume in the sequel to coincide with the inception time T_0, can be expressed as

$$P^{Swn}(t; T_0, T_n, R) = p(t, T_0) \sum_{k=1}^{n} \delta_k E^{Q^{T_0}} \left\{ p(T_0, T_k) \left(R(T_0; T_0, T_n) - R \right)^+ | \mathscr{F}_t \right\}$$

(3.89)

where $R(T_0; T_0, T_n)$ is the swap rate, evaluated at T_0, which for the three model types of this chapter is given by (3.64) when setting $t = T_0$. Since the expression for $R(T_0; T_0, T_n)$ is essentially equivalent in the three types of models, namely those in Sect. 3.2 and that in Sect. 3.3 (where we treat in fact two models together as one type), only the dynamics of $G(t; T_{k-1}, T_k)$, $\frac{p^{\Delta}(t, T_{k-1})}{p^{\Delta}(t, T_k)}$ and $\bar{\nu}_{t, T_{k-1}}$ vary according to the chosen model type, we shall derive the various details for the pricing in the case of the model for $G(t; T_{k-1}, T_k)$ and at the end of the section point out the changes when passing to the other models. Assuming in the sequel $\delta_k = \delta$, for all k, and substituting the first of the relations (3.64) into (3.89), we obtain

$$P^{Swn}(t; T_0, T_n, K) = p(t, T_0) E^{Q^{T_0}} \left\{ \left(\sum_{k=1}^{n} p(T_0, T_k) G(T_0; T_{k-1}, T_k) \right. \right.$$

$$\left. \left. - R\delta \sum_{k=1}^{n} p(T_0, T_k) \right)^+ | \mathscr{F}_t \right\}$$

(3.90)

Recall now that $G(t; T_{k-1}, T_k)$ is a Q^{T_k}-martingale and its dynamics is given under the measure Q by (3.15), while $p(t, T_k)$ corresponds to (3.56).

Since the swaption price has to be computed as an expectation under the forward measure Q^{T_0}, we rewrite (3.56) and (3.15) in terms of the Q^{T_0}-Wiener process w^{T_0} (see (3.17)), thus obtaining

$$p(T_0, T_k) = \frac{p(0, T_k)}{p(0, T_0)} \exp\left[\int_0^{T_0} (A(s, T_0) - A(s, T_k) - \langle \Sigma(s, T_0), \Sigma(s, T_0) - \Sigma(s, T_k) \rangle) ds \right.$$
$$\left. + \int_0^{T_0} (\Sigma(s, T_0) - \Sigma(s, T_k)) dw_s^{T_0} \right]$$
$$G(T_0; T_{k-1}, T_k) = G(0; T_{k-1}, T_k)$$
$$\cdot \exp\left[\int_0^{T_0} (\alpha(s; T_{k-1}, T_k) - \langle \Sigma(s, T_0), \sigma(s; T_{k-1}, T_k) \rangle) ds \right.$$
$$\left. + \int_0^{T_0} \sigma(s; T_{k-1}, T_k) dw_s^{T_0} \right]$$

(3.91)

where $A(s, T_0)$ and $A(s, T_k)$ have to satisfy the drift condition (3.12) and $\alpha(s; T_{k-1}, T_k)$ that in (3.16). To proceed, we make now an additional assumption on the volatilities $\Sigma(s, T)$ and $\sigma(s; T_{k-1}, T_k)$.

Assumption 3.2 Assume that the volatilities have the following structure:

1. Firstly, let

$$
\begin{aligned}
\Sigma(t, T) &= \left[\Sigma^1(t, T), \Sigma^2(t, T) \right] \\
\sigma(t, T, S) &= \left[\sigma^1(t, T, S), \sigma^2(t, T, S) \right]
\end{aligned}
\tag{3.92}
$$

with $\Sigma^1(t, T), \sigma^1(t, T, S)$ being d_1−subvectors, and $\Sigma^2(t, T), \sigma^2(t, T, S)$ corresponding to d_2−subvectors, where $d_1 + d_2 = d$. Furthermore, we assume that $\Sigma^2(t, T) = 0$ and $\sigma^1(t, T, S) = \Sigma^1(t, S) - \Sigma^1(t, T)$.

2. Moreover, the volatilities are separable, i.e.

$$
\begin{aligned}
\Sigma^1(t, T) &= \Lambda(t) \Xi(T), \\
\sigma^2(t, T, S) &= \varsigma(t) \xi(T, S)
\end{aligned}
\tag{3.93}
$$

where $\Lambda : [0, \overline{T}] \to \mathbb{R}_+^{d_1}$ and $\Xi : [0, \overline{T}] \to \mathbb{R}_+$, and $\varsigma : [0, \overline{T}] \to \mathbb{R}_+^{d_2}$ and $\xi : [0, \overline{T}] \times [0, \overline{T}] \to \mathbb{R}_+$.

Note that both assumptions are satisfied for the Vasiček-type volatility structures described in Sect. 3.2.1.2. As we shall see in the sequel, the first assumption will enable us to express the swaption price as an expectation of a function of independent Gaussian variables, similarly to what was done in Sect. 2.5, Corollary 2.4. The second assumption will reduce the total number of variables to two and is often made in HJM models for swaption pricing purposes (cf. Crépey et al. 2015a).

Let now

$$
Z_{T_0} = (Z_{T_0}^1, Z_{T_0}^2) := \left(\int_0^{T_0} \Lambda(s) dw_s^{T_0, d_1}, \int_0^{T_0} \varsigma(s) dw_s^{T_0, d_2} \right)
$$

where w^{T_0, d_1} denotes the first d_1 components and w^{T_0, d_2} the last d_2 components of the Q^{T_0}-Wiener process w^{T_0}. Then

$$
Z_{T_0}^1 \sim N \left(0, \int_0^{T_0} |\Lambda(s)|^2 \, ds \right) =: N(0, \bar{\Gamma}_{T_0}^1)
$$

and

$$
Z_{T_0}^2 \sim N \left(0, \int_0^{T_0} |\varsigma(s)|^2 \, ds \right) =: N(0, \bar{\Gamma}_{T_0}^2)
$$

with $Z_{T_0}^1$ and $Z_{T_0}^2$ independent. Moreover, we have

$$
\begin{aligned}
p(T_0, T_k) &= e^{a^k + b^k Z_{T_0}^1} \\
G(T_0; T_{k-1}, T_k) &= e^{\bar{a}^k + \bar{b}^{k,1} Z_{T_0}^1 + \bar{b}^{k,2} Z_{T_0}^2}
\end{aligned}
\tag{3.94}
$$

where

$$a^k = \log \frac{p(0, T_k)}{p(0, T_0)} + \int_0^{T_0} (A(s, T_0) - A(s, T_k) - \langle \Sigma(s, T_0), \Sigma(s, T_0) - \Sigma(s, T_k) \rangle) ds$$

$$b^k = \mathcal{E}(T_0) - \mathcal{E}(T_k)$$

$$\bar{a}^k = \log G(0; T_{k-1}, T_k) + \int_0^{T_0} (\alpha(s; T_{k-1}, T_k) - \langle \Sigma(s, T_0), \sigma(s; T_{k-1}, T_k) \rangle) ds$$

$$\bar{b}^{k,1} = \mathcal{E}(T_k) - \mathcal{E}(T_{k-1})$$

$$\bar{b}^{k,2} = \xi(T_{k-1}, T_k)$$

are deterministic constants, and the swaption price in (3.90) can be expressed as (for simplicity we shall treat the case $t = 0$)

$$
\begin{aligned}
P^{Swn}(0; T_0, T_n, K) &= p(0, T_0) \\
&\cdot E^{Q^{T_0}} \left\{ \left(\sum_{k=1}^n e^{a^k + \bar{a}^k + (b^k + \bar{b}^{k,1}) Z_{T_0}^1 + \bar{b}^{k,2} Z_{T_0}^2} \right. \right. \\
&\quad \left. \left. - R\delta \sum_{k=1}^n e^{a^k + b^k Z_{T_0}^1} \right)^+ \right\} \\
&= p(0, T_0) E^{Q^{T_0}} \left\{ \left(\sum_{k=1}^n e^{\tilde{A}_k + \tilde{B}_k^1 Z_{T_0}^1 + \tilde{B}_k^2 Z_{T_0}^2} \right. \right. \\
&\quad \left. \left. - R\delta \sum_{k=1}^n e^{A_k + B_k^1 Z_{T_0}^1} \right)^+ \right\}
\end{aligned}
\tag{3.95}
$$

with $\tilde{A}_k = a^k + \bar{a}^k$, $\tilde{B}_k^1 = b^k + \bar{b}^{k,1}$, $\tilde{B}_k^2 = \bar{b}^{k,2}$, $A_k = a^k$ and $B_k^1 = b^k$. This allows us to derive an expression for the swaption price in a completely similar way as in Proposition 2.6 and Corollary 2.4. In order to do so, let us denote

$$
\begin{cases}
g(z_1, z_2) = \sum_{k=1}^n e^{\tilde{A}_k} \exp[\tilde{B}_k^1 z_1 + \tilde{B}_k^2 z_2] \\
h(z_1) = R\delta \sum_{k=1}^n e^{A_k} \exp[B_k^1 z_1]
\end{cases}
\tag{3.96}
$$

We have the following analog of Proposition 2.6 together with Corollary 2.4.

Proposition 3.6 *The price at time $t = 0$ of the swaption described above with maturity T_0 can be computed as*

$$
\begin{aligned}
P^{Swn}(0; T_0, T_n, K) &= p(0, T_0) \sum_{k=1}^n e^{\tilde{A}_k} \int_{\mathbb{R}} e^{\tilde{B}_k^1 z_1} \\
&\cdot \left(\int_{\bar{z}_2(z_1)}^{+\infty} \left[e^{\tilde{B}_k^2 z_2} - e^{\tilde{B}_k^2 \bar{z}_2(z_1)} \right] f_2(z_2) dz_2 \right) f_1(z_1) dz_1 \\
&= p(0, T_0) \sum_{k=1}^n e^{\tilde{A}_k} \int_{\mathbb{R}} e^{\tilde{B}_k^1 z_1} \\
&\quad \left[e^{\frac{1}{2}(\tilde{B}_k^2)^2 \bar{\Gamma}_{T_0}^2} N \left(\frac{\tilde{B}_k^2 \bar{\Gamma}_{T_0}^2 - \bar{z}_2(z_1)}{\sqrt{\bar{\Gamma}_{T_0}^2}} \right) - e^{\tilde{B}_k^2 \bar{z}_2(z_1)} N \left(\frac{-\bar{z}_2(z_1)}{\sqrt{\bar{\Gamma}_{T_0}^2}} \right) \right] f_1(z_1) dz_1
\end{aligned}
\tag{3.97}
$$

where $\bar{z}_2(z_1)$ is the unique solution of the equation $g(z_1, z_2) = h(z_1)$ for $z_1, z_2 \in \mathbb{R}$, f_i denotes the density function of the normal variable $Z_{T_0}^i$, for $i = 1, 2$, and N is the cumulative standard Gaussian distribution function.

Proof From the swaption price expression (3.95), it immediately follows that

$$
P^{Swn}(0; T_0, T_n, K) = p(0, T_0) \int_{\mathbb{R}^2} \left(\sum_{k=1}^{n} e^{\tilde{A}_k + \tilde{B}_k^1 z_1 + \tilde{B}_k^2 z_2} - R\delta \sum_{k=1}^{n} e^{A_k + B_k^1 z_1} \right)^+
$$
$$
\times f_1(z_1) f_2(z_2) dz_1 dz_2,
$$

due to the independence of the random variables $Z_{T_0}^1$ and $Z_{T_0}^2$. Now, in complete analogy to the proof of Proposition 2.6, and noticing that $\tilde{B}_k^2 > 0$, for every $k = 1, \ldots, n$
by Assumption 3.2, it follows that for each $z_1 \in \mathbb{R}$, the function $g(z_1, z_2)$ in (3.96) is
monotonically increasing and continuous in $z_2 \in \mathbb{R}$ with $\lim_{z_2 \to -\infty} g(z_1, z_2) = 0$
and $\lim_{z_2 \to +\infty} g(z_1, z_2) = +\infty$. Hence, there exists a unique $\bar{z}_2(z_1)$ for which
$g(z_1, z_2) = h(z_1)$ and $z_2 \geq \bar{z}_2(z_1) \Leftrightarrow g(z_1, z_2) \geq g(z_1, \bar{z}_2(z_1))$, hence we obtain

$$
P^{Swn}(0; T_0, T_n, K) = p(0, T_0) \sum_{k=1}^{n} e^{\tilde{A}_k} \int_{\mathbb{R}} e^{\tilde{B}_k^1 z_1}
$$
$$
\cdot \left(\int_{\bar{z}_2(z_1)}^{+\infty} \left[e^{\tilde{B}_k^2 z_2} - e^{\tilde{B}_k^2 \bar{z}_2(z_1)} \right] f_2(z_2) dz_2 \right) f_1(z_1) dz_1 \tag{3.98}
$$

where f_i denotes the density function of the Gaussian variable $Z_{T_0}^i$, for $i = 1, 2$. The
second equality in (3.97) now follows in a straightforward manner, just as in the
proof of Corollary 2.4.   □

Remark 3.11 Note that for the $p^\Delta(t, T)$-model of Sect. 3.2.2 the same reasoning
as above leads to an analogous result as in Proposition 3.6, taking into account the
dynamics of the ratios $\frac{p^\Delta(t,T)}{p^\Delta(t,T+\Delta)}$ and making a corresponding assumption on the
volatility of these bond prices. By doing so, one ends up again with a swaption price
expression of exactly the same form (3.95) (with different coefficients) and therefore,
an analog of Proposition 3.6 can be derived. The same comment applies also to the
models of Sect. 3.3, in which the dynamics of $\bar{\nu}_{t, T_{k-1}}$ plays the crucial role.

Remark 3.12 By analogy to Remark 3.9 for caplets, one may wonder if a Black-type
formula for the swaption price could be obtained. This would indeed be possible in a
multiple curve version of the swap market model, where the modeling object would
be the swap rate $R(t; T_0, T_n)$ itself and the swaption could then be seen as a call
option on the swap rate, priced under the swap measure, see (1.45). The swaption
price (3.97) obtained in Proposition 3.6 is obviously not of Black-type, since it is
implied by the log-normal dynamics of individual forward Libor rates and the swap
rate is a convex combination of those, with coefficients which are functions of the
OIS rates, see (1.28). This does not yield log-normal dynamics for the swap rate
under the swap measure (similarly as in the log-normal Libor market model versus
the swap market model).

3.5 Adjustment Factors

It was mentioned at the beginning of this chapter that the quantity $\bar{\nu}_{t,T}$ recalled in
(3.4) plays a basic role in the multi-curve framework: it is crucial for linear derivative pricing in short-rate models (see Sect. 2.3) and HJM instantaneous forward rate
models such as those of Sect. 3.3; furthermore, we also have the relation (3.6) linking
this quantity with $G(t; T, T + \Delta)$. An analogous role is played in the single-curve
framework by the corresponding quantity $\nu_{t,T}$ given by $\nu_{t,T} = \frac{p(t,T)}{p(t,T+\Delta)}$ (see (2.56)).
It may thus be very convenient to have an easy way to pass from $\nu_{t,T}$ to $\bar{\nu}_{t,T}$ by means
of an "adjustment factor". Such a factor has already been derived in Sect. 2.3.2 for
the short-rate setup of Chap. 2. Always in Sect. 2.3.2 it was shown that the adjustment
factor can also be used as a factor to pass directly from a single-curve FRA rate to
a corresponding multi-curve FRA rate. The basic relationship, derived in Chap. 2, is
of the form (see (2.58))

$$\bar{\nu}_{t,T} = \nu_{t,T} \cdot Ad_t^{T,\Delta} \cdot Res_t^{T,\Delta} \tag{3.99}$$

and here we shall derive an analogous relationship also in the context of the models
of this chapter, more precisely those of Sect. 3.3, as well as the $p^\Delta(t, T)$-model of
Sect. 3.2.2, where $\bar{\nu}_{t,T}$ is simply given by $\bar{\nu}_{t,T} = \frac{p^\Delta(t,T)}{p^\Delta(t,T+\Delta)}$.

3.5.1 Adjustment Factor for the Instantaneous Forward Rate Models

In the following proposition we present now a result, completely analogous to Proposition 2.2 in Chap. 2, which serves our original purpose here. The result is stated for
the model $A(t, T)$, $\Sigma(t, T)$, $\bar{A}(t, T)$, $\bar{\Sigma}(t, T)$, but it can be derived in the exact same
manner also for the models $A^f(t, T)$, $\Sigma^f(t, T)$ and $A^\Delta(t, T)$, $\Sigma^\Delta(t, T)$. We have
in fact

Proposition 3.7 *The following relation holds*

$$\bar{\nu}_{t,T} = \nu_{t,T} \cdot Ad_t^{T,\Delta} \cdot Res_t^{T,\Delta} \tag{3.100}$$

where $Ad_t^{T,\Delta}$ will be called "adjustment factor" (see also (2.59)) and it is given by

$$\begin{aligned} Ad_t^{T,\Delta} &:= E^Q\left\{ \frac{p(T,T+\Delta)}{\bar{p}(T,T+\Delta)} \mid \mathscr{F}_t \right\} \\ &= c_{T,T+\Delta} \exp\left[\int_0^t (\Sigma^*(s, T + \Delta) - \Sigma^*(s, T))dw_s \right] \\ &\quad \cdot \exp\left[\frac{1}{2} \int_t^T |\Sigma^*(s, T + \Delta) - \Sigma^*(s, T)|^2 ds \right] \end{aligned} \tag{3.101}$$

with

$$c_{T,T+\Delta} := \frac{p(0,T+\Delta)}{p(0,T)} \frac{\bar{p}(0,T)}{\bar{p}(0,T+\Delta)}$$
$$\cdot \exp\left[\int_0^T (A^*(s, T + \Delta) - A^*(s, T))ds \right] \tag{3.102}$$

The factor $Res_t^{T,\Delta}$ is a residual factor that in some cases is equal to 1 (see e.g. Sect. 3.5.1.1 below). It is given by

$$Res_t^{T,\Delta} := \exp\left[\frac{1}{2} \int_t^T \left(|\Sigma^*(s, T + \Delta) - \Sigma^*(s, T) - \Sigma(s, T)|^2 \right. \right.$$
$$\left. \left. - |\Sigma^*(s, T + \Delta) - \Sigma^*(s, T)|^2 - |\Sigma(s, T)|^2 \right) ds \right] \tag{3.103}$$

$$= \exp\left[- \int_t^T \langle \Sigma(s, T), \Sigma^*(s, T + \Delta) - \Sigma^*(s, T) \rangle \, ds \right]$$

The proof of this proposition consists in a simple verification that, with the expressions for $\nu_{t,T}$ and $\bar{\nu}_{t,T}$ in (3.58) and (3.60) respectively and that of $Ad_t^{T,\Delta}$ given by the rightmost term in (3.101) one obtains for $Res_t^{T,\Delta} := \frac{\bar{\nu}_{t,T}}{\nu_{t,T}}(Ad_t^{T,\Delta})^{-1}$ the expression on the right-hand side of (3.103). For this purpose one has also to use the drift condition on $A(t, T)$ expressed by (3.12) noticing that, by definition, we have $A^*(t, T) = \bar{A}(t, T) - A(t, T)$ and $\Sigma^*(t, T) = \bar{\Sigma}(t, T) - \Sigma(t, T)$. Always with these meanings of $A^*(t, T)$ and $\Sigma^*(t, T)$, the rightmost expression in (3.101) follows from the definition of $Ad_t^{T,\Delta}$ as $Ad_t^{T,\Delta} := E^Q \left\{ \frac{p(T,T+\Delta)}{\bar{p}(T,T+\Delta)} \mid \mathscr{F}_t \right\}$ by standard Itô calculus.

3.5.1.1 Vasiček-Type Volatility Structure

We show here that for the Vasiček example of Sect. 3.2.2.2, the adjustment factor can be given an explicit expression in terms of the model parameters and that the corresponding residual factor is $Res_t^{T,\Delta} = 1$.

With $\sigma(t, T)$ and $\sigma^*(t, T)$ as in (3.37) we have in fact, (see (3.38))

$$\Sigma(t, T) = \left(\frac{\sigma}{b} \left(1 - e^{-b(T-t)} \right), 0 \right) \tag{3.104}$$

and, analogously,

$$\Sigma^*(t, T) = \left(0, \frac{\sigma^*}{b^*} \left(1 - e^{-b^*(T-t)} \right) \right) \tag{3.105}$$

which implies that, on the basis of (3.101),

$$Ad_t^{T,\Delta} = c_{T,T+\Delta} \exp\left[\int_0^t \frac{\sigma^*}{b^*} \left(e^{-b^*(T-s)} - e^{-b^*(T+\Delta-s)} \right) dw_s^2 \right]$$
$$\cdot \exp\left[\frac{1}{2} \int_t^T \left(\frac{\sigma^*}{b^*} \right)^2 \left(e^{-b^*(T-s)} - e^{-b^*(T+\Delta-s)} \right)^2 ds \right] \tag{3.106}$$

where, in the expression (3.102) for $c_{T,T+\Delta}$, the function $A^*(t, T) := \int_t^T a^*(t, u)du$ is a function such that

$$A^*(t, T + \Delta) - A^*(t, T) = \bar{A}(t, T + \Delta) - \bar{A}(t, T) - (A(t, T + \Delta) - A(t, T)) \tag{3.107}$$

where $A(t, T)$ has to satisfy the drift condition in (3.12) and $\bar{A}(t, T)$ has to be compatible with the no-arbitrage condition (3.45) (condition (3.48) for $A^f(t, T)$, respectively condition (3.33) for $A^\Delta(t, T)$).

Concerning, finally, the residual factor $Res_t^{T,\Delta}$, notice that the structure of $\Sigma(t, T)$ and $\Sigma^*(t, T)$ according to (3.104) and (3.105) implies that

$$\langle \Sigma(t, T + \Delta), \Sigma^*(t, T + \Delta) - \Sigma^*(t, T) \rangle = 0$$

so that, by (3.103), $Res_t^{T,\Delta} = 1$.

3.5.2 Adjustment Factor for the HJM-LMM Forward Rate Model

Recall that the adjustment factor was intended to be a factor to pass directly from $\nu_{t,T}$ to $\bar{\nu}_{t,T}$. As in the previous Sect. 3.5.1, we shall start also here by deriving explicit expressions for $\nu_{t,T}$ and $\bar{\nu}_{t,T}$, this time in the context of the HJM-LMM forward rate model, namely the model (3.15) for the processes $G(\cdot; T, T + \Delta)$. Recall that in this model we still consider, as before, the OIS bonds, defined according to (3.14) and having the explicit expression in (3.56), where $A(t, T)$ is supposed to satisfy the drift condition (3.12). On the other hand, instead of the fictitious bonds, we consider here directly the processes $G(\cdot; T, T + \Delta)$ satisfying (3.15) with $\alpha(t; T, T + \Delta)$ subject to the drift condition (3.16).

It follows that for $\nu_{t,T}$ we have the same expression as in the instantaneous forward rate model, namely (3.58) that we reproduce here for convenience

$$\nu_{t,T} = \frac{p(t,T)}{p(t,T+\Delta)}$$
$$= \frac{p(0,T)}{p(0,T+\Delta)} \exp\left[\int_0^t (|\Sigma(s, T + \Delta)|^2 - |\Sigma(s, T)|^2)ds + \int_0^t (\Sigma(s, T + \Delta) - \Sigma(s, T))dw_s \right]$$

$$\tag{3.108}$$

On the other hand, for $\bar{\nu}_{t,T}$ we simply have the relation (see (3.6))

$$\bar{\nu}_{t,T} = G(t; T, T + \Delta) + 1 \tag{3.109}$$

To obtain also here an adjustment factor, we start from working out the ratio $\frac{\bar{\nu}_{t,T}}{\nu_{t,T}}$ on the basis of the above expressions for $\bar{\nu}_{t,T}$ and $\nu_{t,T}$. It turns out that, for this purpose, the direct relation (3.109) is not very convenient but that a more convenient form can be obtained by first differentiating (3.109) thereby taking into account the drift condition (3.16) and then integrating. Defining the shorthand symbol

$$g(t) := \frac{G(t; T, T + \Delta)}{G(t; T, T + \Delta) + 1}$$

this leads to (see also (3.18) with (3.17))

$$
\begin{aligned}
d\bar{\nu}_{t,T} &= d\, G(t; T, T + \Delta) \\
&= G(t; T, T + \Delta)\left[(\alpha(t; T, T + \Delta) + \tfrac{1}{2}|\sigma(t; T, T + \Delta)|^2)dt + \sigma(t; T, T + \Delta)dw_t\right] \\
&= \bar{\nu}_{t,T}\left[(g(t)\alpha(t; T, T + \Delta) + \tfrac{1}{2}g(t)|\sigma(t; T, T + \Delta)|^2)\, dt + g(t)\sigma(t; T, T + \Delta)dw_t\right] \\
&= \bar{\nu}_{t,T}\left[g(t)\langle\sigma(t; T, T + \Delta), \Sigma(t, T + \Delta)\rangle dt + g(t)\sigma(t; T, T + \Delta)dw_t\right]
\end{aligned}
$$
(3.110)

where in the last equality we have taken the drift condition (3.16) into account. Integration now leads to

$$
\bar{\nu}_{t,T} = \bar{\nu}_{0,T}\exp\left[\int_0^t \left(g(s)\langle\sigma(s; T, T + \Delta), \Sigma(s, T + \Delta)\rangle - \tfrac{1}{2}g(s)^2|\sigma(s; T, T + \Delta)|^2\right) ds + \int_0^t g(s)\sigma(s; T, T + \Delta)dw_s\right]
$$
(3.111)

which, together with (3.108), gives us in turn the adjustment factor in the form

$$
\begin{aligned}
\frac{\bar{\nu}_{t,T}}{\nu_{t,T}} = \frac{\bar{\nu}_{0,T}\,p(0,T+\Delta)}{p(0,T)}\exp\Big[&\int_0^t \big(g(s)\langle\sigma(\cdot), \Sigma(s, T + \Delta)\rangle - \tfrac{1}{2}g(s)^2|\sigma(\cdot)|^2 \\
&+ \tfrac{1}{2}\left(|\Sigma(s, T)|^2 - |\Sigma(s, T + \Delta)|^2\right)\big) ds\Big] \\
&\cdot \exp\left[\int_0^t (g(s)\sigma(\cdot) + \Sigma(s, T) - \Sigma(s, T + \Delta))\, dw_s\right]
\end{aligned}
$$
(3.112)

where $(\cdot)$ stands for $(s; T, T + \Delta)$. Note that this corresponds to the multiplicative spread in Sect. 7.3.2 of Henrard (2014).

Chapter 4
Multiple Curve Extensions of Libor Market Models (LMM)

This chapter deals with multiple curve models on a discrete tenor in the spirit of the Libor market models (LMM) and, somewhat differently from the previous two Chaps. 2 and 3, we present here basically just an overview of the major existing approaches.

The Libor market models that were originated by Miltersen et al. (1997) and Brace et al. (1997), were later further developed in several works by Mercurio and co-authors, as well as authors related to them. Consequently these authors were also among the first ones to extend the LMMs to a multi-curve setting. Starting from papers like Morini (2009) and Bianchetti (2010), where the latter uses the analogy with cross-currency modeling to develop a two-curve interest rate model, a series of papers have appeared extending the LMMs to a multi-curve setting, among them Mercurio (2009, 2010a, b, c), Mercurio and Xie (2012) and Ametrano and Bianchetti (2013). This series of papers, in particular Mercurio (2010a) and Mercurio and Xie (2012), which include the developments contained in previous papers authored/ co-authored by Mercurio, form the first approach of which we give an overview in Sect. 4.1. We do not, however, enter into the details of the pricing formulas and the calibration examples, for which we therefore refer to the original papers. Related to the papers by Mercurio (2010a) and Mercurio and Xie (2012) is the paper Ametrano and Bianchetti (2013), where the authors deal in particular with the bootstrapping of various multiple-tenor yield curves, thereby considering essentially only linear interest rate derivatives; here too we simply refer to the original paper.

The other approach, that is alternative to the one just mentioned and that proposes a further theoretical development, is the one in Grbac et al. (2014) which concerns an affine Libor model with multiple curves. A description of this further approach is presented in Sect. 4.2.

The above papers concern mainly "clean valuation" approaches. A more comprehensive, multi-currency, multi-curve approach has been initiated in Fujii et al. (2010, 2011), see also Piterbarg (2010). In parallel, Henrard has developed a more practically oriented approach, for which we refer to his recent book Henrard (2014) that

© The Author(s) 2015
Z. Grbac and W.J. Runggaldier, *Interest Rate Modeling: Post-Crisis Challenges and Approaches*, SpringerBriefs in Quantitative Finance, DOI 10.1007/978-3-319-25385-5_4

synthesizes his previous production, but it is not exactly in the context of an LMM that is our main concern in this chapter. On the other hand Henrard has considered, see e.g. Henrard (2010), multiplicative spreads as example of a form of spreads that are alternative to the additive spreads considered in the above papers and that may turn out to be advantageous in some situations. Multiplicative spreads form also the basis of the approach presented in the paper Cuchiero et al. (2015). Section 4.3 contains a brief overview of the approach in Henrard (2010), as well as that in Cuchiero et al. (2015).

4.1   Multi-curve Extended LMM

As mentioned above, this section is essentially a synthesis of work done by Mercurio and related authors and partly also by Ametrano and Bianchetti (2013).

The classical Libor market models are based on the joint evolution of consecutive forward Libor rates corresponding to a given tenor structure. We recall from the discussion in Chap. 1 that in the classical setup the Libor rate $L(t; T, S)$ was assumed to coincide with the corresponding forward OIS rate $F(t; T, S)$, but this assumption is no longer valid after the crisis.

Recalling the definition of the forward OIS rate in (1.16),

$$F(t; T, S) = E^{Q^S} \{F(T; T, S) \mid \mathscr{F}_t\} = \frac{1}{S - T} \left(\frac{p(t, T)}{p(t, S)} - 1 \right) \quad (4.1)$$

notice that this rate is directly related to the OIS bond prices $p(t, T)$, namely to the discount curve and so the first issue concerns the proper modeling of the discount curve, which occasionally we shall also denote by $p_D(t, T)$. Notice, furthermore, that the discount curve intervenes also in other situations, for example we have that

(i) Swap rates can be represented as linear combinations of forward Libor rates, which are referred to as FRA rates in Mercurio (2010a), with coefficients that depend solely on the discount curve (cf. 1.28).
(ii) Pricing measures correspond to numéraires given by portfolios of OIS bonds and affect thus the drift correction in a measure change.

As already discussed in Chap. 1, a general choice of the discount curve is the OIS curve and we shall do so here as well. This choice is supported by various arguments (recall also Sect. 1.3.1). In particular, collaterals in cash are revalued daily at a rate equal or close to the overnight rate. Note, however, that collaterals can be based also on bonds or other assets, such as foreign currency. In the latter case appropriate adjustments have to be performed for the remuneration rate (see Fujii et al. 2010). The OIS curve is commonly accepted as discount curve and for possible situations, such as in exotics or different currencies, where different discounting is adopted, it can still be considered as a good proxy of the risk-free rate.

Remark 4.1 In this chapter the standing assumption will be that the OIS bonds are tradable assets, see for example Mercurio (2010a) and the comments in Sect. 1.3.1. Note that in the spirit of the Libor market models we do not assume the existence of the OIS short rate r derived from the OIS bond prices and the related martingale measure Q as in the previous chapters, but instead we work directly under the forward measures using the OIS bonds as numéraires.

Since the forward rate $F(t; T, S)$ can also be considered as the fair fixed rate at time $t \le T$ of a forward rate agreement, where the floating rate received at S is $F(T; T, S)$, we shall call $F(t; T, S)$ the forward OIS curve (recall that in practice it can be stripped from OIS swap rates; see also the middle part of Remark 1.2).

Concerning the Libor rates, in the work by Mercurio and related authors an FRA rate is considered that is given as the fair fixed rate at $t \le T$ to be exchanged at time S for the Libor rate $L(T; T, S)$, namely such that this swap has zero value at $t \le T$. Denoting by Q^S the S-forward measure with numéraire $p(t, S)$, the FRA rate is then given by

$$FRA(t; T, S) = E^{Q^S}\{L(T; T, S) \mid \mathscr{F}_t\} \qquad (4.2)$$

where E^{Q^S} denotes expectation with respect to Q^S. This definition, that corresponds to the forward Libor rate $L(t; T, S)$ as in Definition 1.2, has the following advantages

(i) The rates $FRA(t; T, S)$ coincide with the corresponding spot Libor rates at their reset times; they can thus generate any payoff depending on the Libor rates.
(ii) The rates $FRA(t; T, S)$ are martingales under the corresponding forward measures.
(iii) The fact that swap rates can be written as linear combinations of FRA rates with coefficients depending solely on the discount curve is convenient for bootstrapping purposes (see Mercurio 2010a).

In the sequel we shall continue consistently using the name forward Libor rates, keeping in mind that these are by definition the same as the FRA rates from Mercurio (2010a).

4.1.1 Description of the Model

According to a practice followed in the post-crisis setting, the forward Libor rate is mostly viewed as a sum of the forward OIS rate plus a basis/spread (thereby thinking of this basis as a factor driving the Libors in conjunction with the OIS curve). In line with this practice, in the more recent work of Mercurio and related authors an additive spread between the Libor and the OIS curves is considered.

We start with some notation keeping it in line with the above-mentioned papers. As in Sect. 1.3, for given a tenor x, let $\mathscr{T}^x = \{0 \le T_0^x < \cdots < T_{M_x}^x\}$ be a tenor structure compatible with x and denote by δ_k^x the year fraction of the length of the generic kth interval $(T_{k-1}^x, T_k^x]$. Denote by $p(t, T_k^x)$ the price of the OIS bond maturing at T_k^x (discount curve) and set (see (1.16) or (4.1))

$$F_k^x(t) := F(t; T_{k-1}^x, T_k^x) = \frac{1}{\delta_k^x} \left[\frac{p(t, T_{k-1}^x)}{p(t, T_k^x)} - 1 \right] \tag{4.3}$$

Furthermore, as mentioned at the end of the previous subsection, since the FRA rates, as given in (4.2), were introduced in relation to the FRAs with the underlying Libor rates, set

$$L_k^x(t) := FRA(t; T_{k-1}^x, T_k^x) \tag{4.4}$$

and call $L_k^x(t)$ the forward Libor rate.

Notice that the FRA rates as introduced in (4.2) correspond to what is called a standard (or text-book) FRA. This has to be contrasted with the so-called "market FRA", cf. Remark 1.3, which differ from the standard ones in that the payment is made at the beginning of the reference interval, discounted by the corresponding Libor rate. Hence, taking $T = T_{k-1}^x$, $S = T_k^x$ for a generic k, one has

$$P^{mFRA}(T_{k-1}^x; T_{k-1}^x, T_k^x, R, 1) = \frac{\delta_k^x \left(L(T_{k-1}^x; T_{k-1}^x, T_k^x) - R \right)}{1 + \delta_k^x L(T_{k-1}^x; T_{k-1}^x, T_k^x)} \tag{4.5}$$

implying that (see Appendix A in Mercurio 2010b), at $t < T_{k-1}^x$ one has,

$$R^{mFRA}(t; T_{k-1}^x, T_k^x) = \frac{1}{\delta_k^x} \left[\frac{1}{E^{Q^{T_{k-1}^x}} \left\{ \frac{1}{1 + \delta_k^x L(T_{k-1}^x; T_{k-1}^x, T_k^x)} \mid \mathscr{F}_t \right\}} - 1 \right] \tag{4.6}$$

As recalled in Remark 1.3, Mercurio (2010b) (see also Ametrano and Bianchetti 2013) points out that the difference in the values $P^{mFRA}(T_{k-1}^x; T_{k-1}^x, T_k^x, R, 1)$ and $P^{FRA}(T_{k-1}^x; T_{k-1}^x, T_k^x, R, 1)$ is generally small, so that one can limit oneself to standard FRA rates also for what concerns a possible bootstrapping from market FRA rates.

As already mentioned, following the practice to build Libor curves at a spread over the OIS curve, Mercurio and the related authors consider additive spreads that can now be defined as

$$S_k^x(t) := L_k^x(t) - F_k^x(t) \tag{4.7}$$

Additive spreads have the advantage that, since L_k^x and F_k^x are martingales under $Q^{T_k^x}$, so is also S_k^x. To model their dynamics under $Q^{T_k^x}$ one thus needs to specify only the volatility and correlation structure.

Having now the three quantities $L_k^x(t)$, $F_k^x(t)$, $S_k^x(t)$, we need to introduce dynamics for them. Given the relationship (4.7), we need only the joint dynamics of two of the three quantities, which then induces the dynamics also for the third one. The aim thereby should be to achieve model tractability in view of interest rate derivative pricing, as well as a convenient setup for calibration and bootstrapping. In Mercurio (2010a, b) the author chooses to jointly model $F_k^x(t)$ and $S_k^x(t)$ that has as main advantage the direct modeling of the spread allowing thus to model its dynamics so that it remains positive. Such a choice is made also in Fujii et al. (2011). Notice furthermore that when, as required for a multi-curve setup, one has to model multiple tenors simultaneously, i.e. $F_k^{x_i}(t)$ and $S_k^{x_i}(t)$ for different values x_i of the tenor x, one has to properly account for possible no-arbitrage relations that have to hold across different time intervals. To this effect notice that only forward OIS rates with different tenors are constrained by no-arbitrage relations; the associated spreads are relatively free to move independently from one another. Given this freedom, one may try to derive models that preserve the tractability of the single tenor case, especially in view of pricing optional derivatives in closed form. In this sense, in Mercurio (2010a, b) an approach is presented by choosing the dynamics of the OIS rates and related spreads so that they are similar for all considered tenors. Furthermore, as the title in Mercurio and Xie (2012) puts it, the spread should by all means be stochastic. In fact, a relatively simple approach would be to elect a given forward OIS curve as reference curve and model all other curves at a deterministic spread over the reference curve. However, this is in contrast with the empirical evidence (see Figs. 1.3 (left) and 1.4) and furthermore, with a deterministic basis, the Libor-OIS swaption price would in some situations, e.g. OTM swaptions, turn out to be zero. Although the impact of a stochastic basis on the pricing of exotic products is difficult to assess a priori, in Mercurio and Xie (2012) it is shown that also with a suitably modeled stochastic basis one may achieve very good tractability.

4.1.2 Model Specifications

We mention here specific models considered in Mercurio (2010b) and Mercurio and Xie (2012) where, in line with the classical LMMs, log-normal and shifted log-normal models are considered, but with stochastic volatility in form of Heston or SABR.

For the multi-curve setup consider now, as in Sect. 1.3, different possible tenor values $x_1 < x_2 < \cdots < x_n$ and the associated tenor structures $\mathscr{T}^{x_i} = \{0 \le T_0^{x_i} < \cdots < T_{M_{x_i}}^{x_i}\}$, thereby assuming that $\mathscr{T}^{x_n} \subset \mathscr{T}^{x_{n-1}} \subset \cdots \subset \mathscr{T}^{x_1} \subseteq \mathscr{T}$.

Denote then by $L_k^{x_i}(t)$, $F_k^{x_i}(t)$, $S_k^{x_i}(t)$ the corresponding relevant quantities and let $\delta_k^{x_i}$ be the year fraction of $T_k^{x_i} - T_{k-1}^{x_i}$.

Starting from the forward OIS rates, in Mercurio (2010b) these are modeled according to the following shifted-type dynamics

$$dF_k^{x_i}(t) = \sigma_k^{x_i}(t) V^F(t) \left[\frac{1}{\delta_k^{x_i}} + F_k^{x_i}(t) \right] dZ_k^{F,x_i}(t) \qquad (4.8)$$

where $\sigma_k^{x_i}$ are deterministic functions and Z_k^{F,x_i} are Wiener processes under the forward measures $Q^{T_k^{x_i}}$, for all $k = 1, \ldots, M_{x_i}$. The process $V^F(t)$ is a common factor process with $V^F(0) = 1$ and independent of all Z_k^{F,x_i}. As mentioned previously, the $F_k^{x_i}$ have to satisfy no-arbitrage consistency conditions and in Mercurio (2010b) it is shown that they are given by

$$\sigma_k^{x_i}(t) = \sum_{j=i_{k-1}+1}^{i_k} \sigma_j^{x_1}(t) \tag{4.9}$$

namely the volatility coefficient $\sigma_k^{x_i}$ of $F_k^{x_i}$ has to be equal to the sum of the volatility coefficients $\sigma_j^{x_1}$ of the rates $F_j^{x_1}$ for $j \in \{i_{k-1}+1, i_{k-1}+2, \ldots, i_k\}$. Here j correspond to the indices of the tenor dates $T_{k-1}^{x_i} = T_{i_{k-1}}^{x_1} < T_{i_{k-1}+1}^{x_1} < \cdots < T_{i_k}^{x_1} = T_k^{x_i}$ of the tenor structure $\mathcal{T}^{x_1}$ falling between the dates $T_{k-1}^{x_i}$ and $T_k^{x_i}$ of the tenor structure $\mathcal{T}^{x_i}$.

Coming now to the Libor-OIS spreads, Mercurio and Xie (2012) start from the following general model

$$S_k^{x_i}(t) = \phi_k^{x_i}(F_k^{x_i}(t), X_k^{x_i}(t)) \tag{4.10}$$

where $X_k^{x_i}$ are factor processes and the functions $\phi_k^{x_i}$ have to be chosen so that $S_k^{x_i}$ are martingales under $Q^{T_k^{x_i}}$. In particular, in Mercurio and Xie (2012) the functions $\phi_k^{x_i}$ are chosen to be affine functions with the advantage that the parameters in the affine specification can be explained in terms of correlations between OIS rates and basis spreads, as well as in terms of their variances. If also the forward OIS rates follow a convenient model, then caplets and swaptions can be priced in semi-closed form.

As for the factor processes $X_k^{x_i}$, things can be simplified without too much loss of flexibility by taking them to be independent of k, i.e. $X_k^{x_i}(t) = X^{x_i}(t)$, for all k, with $X^{x_i}(t)$ following a log-normal model. If, then, the OIS rate is, say, of the $G1^{++}$ form (i.e. one-factor Hull and White (1990) model with deterministic shift to be calibrated to the initial term structure), with the affine model choice for $\phi_k^{x_i}$ one obtains semi-analytic pricing formulas for caplets and swaptions in the sense that what is required is a one-dimensional integration of a closed-form function of the Black-Scholes type (see Mercurio and Xie 2012 for the case of a swaption).

Remaining always with the Libor-OIS spreads, Mercurio (2010b) assumes more specifically a model of the form

$$S_k^{x_i}(t) = S_k^{x_i}(0) M^{x_i}(t), \quad k = 1, \ldots, M_{x_i} \tag{4.11}$$

where, analogously to the previous case where $X_k^{x_i}(t) = X^{x_i}(t)$, also M^{x_i} remains the same for all k and is defined by the following SABR-type process

$$\begin{cases} dM^{x_i}(t) = (M^{x_i}(t))^{\beta^{x_i}} V^{x_i}(t) \, dZ^{x_i}(t) \\[2mm] dV^{x_i}(t) = \varepsilon^{x_i} V^{x_i}(t) \, dW^{x_i}(t) \end{cases} \tag{4.12}$$

with $\beta^{x_i} \in (0, 1]$, $\varepsilon^{x_i} > 0$ and where Z^{x_i} and W^{x_i} are Wiener processes with respect to each forward measure $Q^{T_k^{x_i}}$, independent of the Wiener processes Z_k^{F, x_i} in (4.8), but which may be correlated, i.e. $dZ^{x_i} dW^{x_i} = \rho^{x_i} dt$ with $\rho^{x_i} \in [-1, 1]$. This model allows for convenient caplet and, in some particular cases, swaption pricing (see Mercurio 2010b). The fact that the rates and the spreads with different tenors x_i follow the same type of dynamics, leads to similar pricing formulas for caps and swaptions even if they are based on different tenors. This is particularly convenient for simultaneous option pricing across different tenors, as well as for calibration.

4.2 Affine Libor Models with Multiple Curves

In this section we present the model developed in Grbac et al. (2014), which is also a discrete-tenor model and is based on affine driving processes. This modeling approach has first been proposed in Keller-Ressel et al. (2013) in the single-curve case and further extended in Grbac et al. (2014) to the multiple curve setup. The main advantage of this framework is its ability to ensure positive interest rates and spreads by construction, and at the same time, produce semi-analytic caplet and swaption pricing formulas. In contrast to Mercurio (2010b), the OIS rates F_k^x and the Libor rates L_k^x are chosen as modeling quantities, which ensures straightforward pricing of caplets and at the same time the positivity of spreads S_k^x can be easily obtained. These features are due to a specific model construction, which relies on a family of parametrized martingales greater or equal to one and increasing with respect to the parameter. When such martingales are taken as building blocks of the model, as we shall see below, the positivity of interest rates and spreads follows simply by construction. The second important point is that affine processes are chosen as driving processes for this family of martingales, thus guaranteeing analytic tractability of the model and, consequently, semi-analytic pricing formulas for non-linear derivatives based on Fourier transform methods.

4.2.1 The Driving Process and Its Properties

In this section we shall fix the probability space and the driving process that we are going to work with. For sake of simplicity, we choose to work with affine diffusions in order to present the model in a concise and simple manner. Another reason is that affine diffusions were already used as driving processes in Chap. 2 and, hence, we can rely on the technical preliminaries from that chapter. Since the model construction requires a positive affine process as a driving process, this boils down to multidimensional CIR processes. However, we emphasize that the original paper of Grbac et al. (2014) is not limited to this class and the model is based on general positive affine processes allowing for jumps as well. This is especially important in view of model calibration, where the additional flexibility coming from the jumps is exploited to ensure a better fit to market data.

Let $(\Omega, \mathscr{F}, \mathbb{F}, P)$ denote a complete stochastic basis, where $\mathbb{F} = (\mathscr{F}_t)_{t \in [0,T]}$ and T denotes some finite time horizon. Consider a stochastic process $X = (X^1, \ldots, X^d)$, where each component X^i solves the SDE

$$dX_t^i = (a^i - b^i X_t^i)dt + \sigma^i \sqrt{X_t^i}\, dw_t^i \qquad (4.13)$$

with w^i a Wiener process such that all Wiener processes $w^i, i = 1, \ldots, d$, are mutually independent. The coefficients a^i, b^i and σ^i are positive and $a^i \geq \frac{(\sigma^i)^2}{2}$.

Denote

$$\mathscr{I}_T := \left\{ u = (u^1, \ldots, u^d) \in \mathbb{R}^d : E_x\left\{ e^{\langle u, X_T\rangle} \right\} < \infty \right\}$$

where E_x denotes the expectation conditional on $X_0 = x$. Note that this is a multi-dimensional analog of the set $\mathscr{I}_T$ defined in (2.12). Then according to Lemma 2.2, the conditional moment generating function of X_T has the following exponentially affine form:

$$E\left\{ \exp\langle u, X_T\rangle \,\big|\, \mathscr{F}_t \right\} = \exp\left(A^u(T - t) + \langle B^u(T - t), X_t\rangle \right) \qquad (4.14)$$

for all $u \in \mathbb{R}_+^d \cap \mathscr{I}_T$ and $0 \leq t \leq T$. Here $A^u(T - t)$ and $B^u(T - t)$ are obtained by applying Lemma 2.2 to each component X^i of X and using independence. This yields

$$A^u(T - t) = \sum_{i=1}^d A^{u,i}(T - t), \qquad B^u(T - t) = (B^{u,1}(T - t), \ldots, B^{u,d}(T - t))$$

where, for each $i = 1, \ldots, d$, $A^{u,i}(T - t)$ and $B^{u,i}(T - t)$ correspond to $A(t, T) = A(T - t)$ and $-B(t, T) = -B(T - t)$ in Lemma 2.2 applied to the process X^i with $\gamma = 0$ and $K = -u$. In Eq. (4.14) $\langle \cdot, \cdot\rangle$ denotes the inner product on $\mathbb{R}^d$.

An essential ingredient in affine Libor models, as we shall see in the next subsection, is the construction of parametrized martingales which are greater than or equal to one and increasing in this parameter, see the review paper by Papapantoleon (2010). The following two lemmas, taken from Grbac et al. (2014) and Keller-Ressel et al. (2013), summarize the main ideas and properties on which the construction will be based.

Lemma 4.1 *Consider the affine process X defined above and let $u \in \mathbb{R}_+^d \cap \mathscr{I}_T$. Then the process $M^u = (M_t^u)_{t\in[0,T]}$ with*

$$M_t^u = E\left\{ e^{\langle u, X_T\rangle} \,|\, \mathscr{F}_t \right\} = \exp\left(A^u(T - t) + \langle B^u(T - t), X_t\rangle \right) \qquad (4.15)$$

is a P-martingale, greater than or equal to one, and the mapping $u \mapsto M_t^u$ is increasing, for every $t \in [0, T]$.

In the lemma below inequalities involving vectors are interpreted componentwise.

Lemma 4.2 *The functions $A^u(t)$ and $B^u(t)$ in (4.14) satisfy the following:*

1. *$A^0(t) = B^0(t) = 0$ for all $t \in [0, T]$.*
2. *For each $t \in [0, T]$, the functions $\mathscr{I}_T \ni u \mapsto A^u(t)$ and $\mathscr{I}_T \ni u \mapsto B^u(t)$ are (componentwise) convex.*

3. $u \mapsto A^u(t)$ and $u \mapsto B^u(t)$ are order-preserving: let $(t, u), (t, v) \in [0, T] \times \mathscr{I}_T$, with $u \leq v$. Then

$$A^u(t) \leq A^v(t) \quad and \quad B^u(t) \leq B^v(t) \tag{4.16}$$

4. $u \mapsto B^u(t)$ is strictly order-preserving: let $(t, u), (t, v) \in [0, T] \times \mathscr{I}_T$, with $u < v$. Then $B^u(t) < B^v(t)$.

4.2.2 The Model

Consider again the tenor structures introduced in Sect. 1.3 and used in Sect. 4.1.2, $\mathscr{T}^{x_n} \subset \mathscr{T}^{x_{n-1}} \subset \cdots \subset \mathscr{T}^{x_1} \subseteq \mathscr{T}$, where $\mathscr{T}^{x_i} = \{0 \leq T_0^{x_i} < \cdots < T_{M_{x_i}}^{x_i}\}$, for each $i = 1, \ldots, n$, and where $\delta_k^{x_i}$ denotes the year fraction of $T_k^{x_i} - T_{k-1}^{x_i}$. Recall that we assume that all $T_{M_{x_i}}^{x_i}$ coincide and denote by T_M the common terminal date for all tenor structures. As earlier, denote by $\mathscr{X} := \{x_1, \ldots, x_n\}$ the set of all tenors and for each tenor $x \in \mathscr{X}$, let $\mathscr{K}^x := \{1, 2, \ldots, M_x\}$ denote the collection of all subscripts related to the tenor structure $\mathscr{T}^x$. In the sequel we shall assume for ease of notation that each tenor structure is equidistant, i.e. $\delta_k^{x_i} = \delta^{x_i}$.

As earlier, we consider the OIS curve as discount curve. We denote by $p(t, T)$ the discount factor, i.e. the price of the OIS bond at time t for maturity T.

Moreover, let Q^{T_M} denote the terminal forward measure, i.e. the martingale measure associated with the numéraire $p(\cdot, T_M)$, which is supposed to be given. The corresponding expectation is denoted by E^{T_M}. Then, we introduce forward measures $Q^{T_k^x}$ associated to the numéraires $p(\cdot, T_k^x)$ for every tenor x and $k \in \mathscr{K}^x$. The corresponding expectation is denoted by $E^{T_k^x}$. The forward measures $Q^{T_k^x}$ are equivalent to Q^{T_M}, and defined in the usual way via

$$\left.\frac{dQ^{T_k^x}}{dQ^{T_M}}\right|_{\mathscr{F}_t} = \frac{p(0, T_M)}{p(0, T_k^x)} \frac{p(t, T_k^x)}{p(t, T_M)} \tag{4.17}$$

As seen in Sect. 4.1, the main modeling objects in the multiple curve LMM setting are the forward OIS rates F_k^x, the forward Libor rates L_k^x and the spreads S_k^x. Recalling again Mercurio (2010b), a good model for the dynamic evolution of the forward OIS and forward Libor rates, and thus also of their spread, should satisfy certain conditions which stem from economic reasoning, arbitrage requirements and the definitions of these rates. Grbac et al. (2014) formulate these conditions as model requirements:

(M1) $F_k^x(t) \geq 0$ and $F_k^x \in \mathscr{M}(Q^{T_k^x})$, for all $x \in \mathscr{X}$, $k \in \mathscr{K}^x$, $t \in [0, T_{k-1}^x]$.
(M2) $L_k^x(t) \geq 0$ and $L_k^x \in \mathscr{M}(Q^{T_k^x})$, for all $x \in \mathscr{X}$, $k \in \mathscr{K}^x$, $t \in [0, T_{k-1}^x]$.
(M3) $S_k^x(t) \geq 0$ and $S_k^x \in \mathscr{M}(Q^{T_k^x})$, for all $x \in \mathscr{X}$, $k \in \mathscr{K}^x$, $t \in [0, T_{k-1}^x]$.

Here $\mathscr{M}(Q^{T_k^x})$ denotes the set of $Q^{T_k^x}$-martingales.

The model presented in the sequel satisfies the above conditions by construction, while producing tractable dynamics for all three processes under all forward measures. The approach was first introduced by Keller-Ressel et al. (2013) and then extended to the multiple curve setup by Grbac et al. (2014). The first step is the construction of two families of parametrized Q^{T_M}-martingales driven by the process X defined in the previous subsection, which is assumed to be affine under the measure Q^{T_M}.

More precisely, assume that the process X starts at the canonical value $\mathbf{1} = (1, \ldots, 1)$ and assume that two sequences of vectors $(u_k)_{k \in \mathbb{N}}$ and $(v_k)_{k \in \mathbb{N}}$ in $\mathbb{R}_+^d \cap \mathscr{I}_T$ are given. Then one constructs two families of parametrized Q^{T_M}-martingales following the method described in Lemma 4.1 by setting

$$M_t^{u_k} = \exp\left(A^{u_k}(T_M - t) + \langle B^{u_k}(T_M - t), X_t \rangle\right) \tag{4.18}$$

and

$$M_t^{v_k} = \exp\left(A^{v_k}(T_M - t) + \langle B^{v_k}(T_M - t), X_t \rangle\right) \tag{4.19}$$

By Lemma 4.1, $M_t^{u_k} \geq 1$ and $M_t^{v_k} \geq 1$, for all k. Note, moreover, that the ordering of parameters u_k and v_k carries over to the martingales related to them; for example, if $u_{k-1} \geq u_k$, then $M_t^{u_{k-1}} \geq M_t^{u_k}$ for all $t \geq 0$. Such families of martingales are then used to model the forward OIS and Libor rates.

Let us fix an arbitrary tenor x and the associated tenor structure $\mathscr{T}^x$. We begin by presenting the model for the OIS rates. This model is completely analogous to the single-curve model introduced by Keller-Ressel et al. (2013). In the first step, one notices that

$$1 + \delta^x F_k^x(t) = \frac{p(t, T_{k-1}^x)}{p(t, T_k^x)} = \frac{\frac{p(t, T_{k-1}^x)}{p(t, T_M)}}{\frac{p(t, T_k^x)}{p(t, T_M)}} \tag{4.20}$$

where the forward price process $\frac{p(\cdot, T_k^x)}{p(\cdot, T_M)}$ is a Q^{T_M}-martingale for any $k \in \mathscr{K}^x$. Therefore, to model the OIS rates F_k^x, one begins by postulating the dynamics of each forward price process $\frac{p(\cdot, T_k^x)}{p(\cdot, T_M)}$:

$$\frac{p(t, T_k^x)}{p(t, T_M)} = M_t^{u_k^x}, \qquad k \in \mathscr{K}^x, \, t \leq T_k^x \tag{4.21}$$

where $u_k^x \in \mathbb{R}_+^d \cap \mathscr{I}_T$ is some vector. The second step follows from (4.20) and (4.21), namely

$$1 + \delta^x F_k^x(t) = \frac{M_t^{u_{k-1}^x}}{M_t^{u_k^x}} \tag{4.22}$$

and we note that this process is a $Q^{T_k^x}$-martingale since $M^{u_{k-1}^x}$ is a Q^{T_M}-martingale and $M^{u_k^x} = \frac{p(\cdot, T_k^x)}{p(\cdot, T_M)}$ is the density process for the measure change from Q^{T_M} to $Q^{T_k^x}$ up to a normalizing constant; compare with (4.17). Finally, the third step relies on the observation that, if the vectors u_k^x are chosen to be (componentwise) decreasing with respect to k, i.e. $u_{k-1}^x \geq u_k^x$, it follows $M^{u_{k-1}^x} \geq M^{u_k^x}$ and therefore $1 + \delta^x F_k^x(t) \geq 1$, or equivalently $F_k^x(t) \geq 0$.

Now on top of the model for the OIS rates, it remains to suitably specify the dynamics of the Libor rates L_k^x in order to completely specify the multiple curve model. To do so, a similar idea can be used. More precisely, one postulates that

$$1 + \delta^x L_k^x(t) = \frac{M_t^{v_{k-1}^x}}{M_t^{u_k^x}} \tag{4.23}$$

for every $k = 2, \ldots, M_x$ and $t \in [0, T_{k-1}^x]$ and where $v_{k-1}^x \in \mathbb{R}_+^d \cap \mathscr{I}_T$ is some vector. Hence, the process $1 + \delta^x L_k^x$ is a $Q^{T_k^x}$-martingale by exactly the same arguments as above. In addition, if $v_{k-1}^x \geq u_k^x$, then $M^{v_{k-1}^x} \geq M^{u_k^x}$ and $1 + \delta^x L_k^x(t) \geq 1$, or equivalently $L_k^x(t) \geq 0$.

This procedure presents the main modeling idea. The questions that still have to be answered are, if such sequences of vectors (u_k^x) and (v_k^x) can be found for any given initial term structure of the forward OIS and the forward Libor rates. Moreover, in case of an affirmative answer, do these sequences possess the desired monotonicity properties? The following proposition, which summarizes the results shown by Grbac et al. (2014), describes the main properties of the model and explains how to construct it from a given initial term structure of OIS and Libor rates.

Proposition 4.1 *Consider the finest tenor structure $\mathscr{T}$, let $p(0, T_l)$, $l \in \mathscr{K}$, be the initial term structure of non-negative OIS bond prices and assume that*

$$p(0, T_1) \geq \cdots \geq p(0, T_M)$$

Moreover, for a fixed tenor x and the corresponding tenor structure $\mathscr{T}^x$, let $L_k^x(0)$, $k \in \mathscr{K}^x$, be the initial term structure of non-negative forward Libor rates and assume that for every $k \in \mathscr{K}^x$

$$L_k^x(0) \geq \frac{1}{\delta^x} \left(\frac{p(0, T_{k-1}^x)}{p(0, T_k^x)} - 1 \right) = F_k^x(0) \tag{4.24}$$

Then the following statements are true:

1. *There exists a decreasing sequence $u_0 \geq u_1 \geq \cdots \geq u_M = 0$ in $\mathbb{R}_+^d \cap \mathscr{I}_T$, such that*

$$M_0^{u_l} = \frac{p(0, T_l)}{p(0, T_M)} \quad \text{for all } l \in \mathscr{K} \tag{4.25}$$

   Furthermore, for each $k \in \mathscr{K}^x$, we set

$$u_k^x := u_l \qquad\qquad (4.26)$$

where $l \in \mathcal{K}$ is such that $T_l = T_k^x$.
2. *There exists a sequence $v_0^x, v_1^x, \ldots, v_{M_x}^x = 0$ in $\mathbb{R}_+^d \cap \mathscr{I}_T$, such that $v_k^x \geq u_k^x$ and*

$$M_0^{v_k^x} = (1 + \delta^x L_{k+1}^x(0)) M_0^{u_{k+1}^x}, \quad for all \; k = 0, 1, \ldots, M_x - 1 \qquad (4.27)$$

3. *If X is one-dimensional, the sequences $(u_k^x)_{k\in\mathcal{K}^x}$ and $(v_k^x)_{k\in\mathcal{K}^x}$ are unique.*
4. *If all initial forward OIS rates and initial spreads are positive, then the sequence (u_k^x) is strictly decreasing and $v_k^x > u_k^x$, for all $k = 0, 1, \ldots, M_x - 1$.*

Therefore, from the model construction it follows immediately:

1. F_k^x and L_k^x are $Q^{T_k^x}$-martingales, for every $k \in \mathcal{K}^x$.
2. $L_k^x(t) \geq F_k^x(t) \geq 0$, for every $k \in \mathcal{K}^x$ and $t \in [0, T_{k-1}^x]$.

Remark 4.2 The results of Proposition 4.1 confirm that, for given initial term structures, the affine Libor model with multiple curves can theoretically be constructed by choosing sequences (u_k^x) and (v_k^x) as described above. Regarding the practical implementation of the model, one notices that when the driving process is multidimensional (which will typically be the case in applications), the vector parameters (u_k^x) and (v_k^x) are not unique and it seems that there is no canonical choice for them. This in turn gives a freedom to devise special cases of the model by pre-choosing various suitable structures for these sequences and then fitting the initial structures. One such example is a factor model with common and idiosyncratic components for each OIS and Libor rate, which is presented in Sect. 8 of Grbac et al. (2014) and where this is achieved by setting some components of (u_k^x) and (v_k^x) to zero or mutually equal in order to exclude the effect of certain components of the driving process on each specific rate.

Remark 4.3 The multiple curve affine Libor model is constructed under the terminal forward measure Q^{T_M}. Grbac et al. (2014) show that the model structure is preserved under different forward measures. More precisely, the process X remains an affine process, although its 'characteristics' become time-dependent, under any forward measure other than Q^{T_M}. The affine property plays a crucial role in the derivation of tractable pricing formulas for interest rate derivatives in the next subsection. Note, furthermore, that the multiple curve affine LIBOR model fulfills requirements (M1)–(M3), which are consistent with the typical market observations of nonnegative interest rates and spreads. However, a phenomenon of negative values of various interest rates has been continually observed in the European markets starting from the second half of 2014 and thus, it is worthwhile mentioning that also negative interest rates can be easily accommodated in this setup by considering, for example, affine processes on $\mathbb{R}^d$ instead of $\mathbb{R}_+^d$ or 'shifted' positive affine processes where supp(X) $\in [a, \infty)^d$ with $a < 0$. A specification of the multiple curve affine Libor model allowing for negative rates and positive spreads is presented in Sect. 4.1 of that paper.

Remark 4.4 (Connection to Libor market models) Having a model of this Sect. 4.2 in which the dynamics of the OIS and the Libor rates are given by (4.22) and (4.23), it is natural to look for a possible relationship between this model and the multiple curve Libor market models of Sect. 4.1. This relationship has been established in Grbac et al. (2014). More precisely, starting from (4.22) and the definition of the martingales $M^{u_k^x}$ given in (4.18), as well as the definition of the process X in (4.13), and using Itô's formula and some algebraic manipulations, one arrives at the dynamics of the OIS rate F_k^x under the forward measure $Q^{T_k^x}$

$$\frac{dF_k^x(t)}{F_k^x(t)} = \Gamma_{x,k}^{\mathsf{T}}(t)\,dw_t^{x,k} \tag{4.28}$$

with the volatility structure $\Gamma_{x,k} = (\Gamma_{x,k}^1, \ldots, \Gamma_{x,k}^d) \in \mathbb{R}_+^d$ provided by

$$\Gamma_{x,k}^i(t) := \frac{1 + \delta^x F_k^x(t)}{\delta^x F_k^x(t)} \left(B^{u_{k-1}^x,i}(T_M - t) - B^{u_k^x,i}(T_M - t) \right) \sqrt{X_t^i}\,\sigma^i \tag{4.29}$$

and the $Q^{T_k^x}$-Wiener process $w^{x,k} = (w^{x,k,1}, \ldots, w^{x,k,d})$ given by

$$
\begin{aligned}
w^{x,k,i} &:= w^i - \sum_{l=k+1}^{M_x} \int_0^\cdot \frac{\delta^x F_l^x(t)}{1 + \delta^x F_l^x(t)} \Gamma_{x,l}^i(t)\,dt \\
&= w^i - \sum_{l=k+1}^{M_x} \int_0^\cdot \left(B^{u_{l-1}^x,i}(T_M - t) - B^{u_l^x,i}(T_M - t) \right) \sqrt{X_t^i}\,\sigma_i\,dt.
\end{aligned}
\tag{4.30}
$$

Note from (4.29) that the volatility structure is determined by σ^i and by the driving process itself via $B^{u_{k-1}^x,i}(T_M - t)$ and $B^{u_k^x,i}(T_M - t)$, and also that there is a built-in shift in the model by construction. Furthermore, we notice that the dynamics of the OIS rates in Eq. (4.28) correspond to (4.8) in the Libor market model of Sect. 4.1.

In complete analogy, starting from the dynamics of the Libor rates in (4.23) and introducing the volatility structure $\Lambda_{x,k} = (\Lambda_{x,k}^1, \ldots, \Lambda_{x,k}^d) \in \mathbb{R}_+^d$

$$\Lambda_{x,k}^i(t) := \frac{1 + \delta^x L_k^x(t)}{\delta^x L_k^x(t)} \left(B^{v_{k-1}^x,i}(T_M - t) - B^{u_k^x,i}(T_M - t) \right) \sqrt{X_t^i}\,\sigma^i \tag{4.31}$$

one obtains for L_k^x the following $Q^{T_k^x}$-dynamics in the spirit of the Libor market model

$$\frac{dL_k^x(t)}{L_k^x(t)} = \Lambda_{x,k}^{\mathsf{T}}(t)\,dw_t^{x,k} \tag{4.32}$$

where $w^{x,k}$ is the $Q^{T_k^x}$-Wiener process given above.

4.2.3   Pricing in the Multiple Curve Affine Libor Model

Thanks to its tractability under all forward measures, the multiple curve affine Libor model allows for semi-analytical pricing of interest rate derivatives. Below we present the main results from Grbac et al. (2014) and refer to the paper for details. In particular, the valuation of caplets based on Fourier transform methods is of the same complexity as the valuation of caplets in the single-curve affine Libor model. In order to obtain semi-closed pricing formulas for swaptions, Grbac et al. (2014) make use of the linear boundary approximation proposed by Singleton and Umantsev (2002) combined with Fourier transform methods.

Let us begin by considering linear derivatives, namely interest rate swaps and basis swaps. We use the definitions and the notation from Sects. 1.4.3 and 1.4.5, assuming for simplicity that the nominal is $N = 1$. Since the modeling objects in the multiple curve affine Libor model are directly the forward Libor rates, it is straightforward to conclude that the time-t value of the interest rate swap on the tenor structure $\mathcal{T}^x$ with fixed rate denoted by R is given by

$$P^{Sw}(t; \mathcal{T}^x, R) = \delta^x \sum_{k=1}^{M_x} p(t, T_k^x) \left(L_k^x(t) - R \right)$$

and the fair swap rate $R(t; \mathcal{T}^x)$ is therefore

$$R(t; \mathcal{T}^x) = \frac{\sum_{k=1}^{M_x} p(t, T_k^x) L_k^x(t)}{\sum_{k=1}^{M_x} p(t, T_k^x)} \tag{4.33}$$

Similarly, the time-t value of the basis swap defined on the tenor structures $\mathcal{T}^{x_1}$ and $\mathcal{T}^{x_2}$ with spread S is expressed as

$$P^{BSw}(t; \mathcal{T}^{x_1}, \mathcal{T}^{x_2}) = \sum_{i=1}^{M_{x_1}} \delta^{x_1} p(t, T_i^{x_1}) L_i^{x_1}(t) - \sum_{j=1}^{M_{x_2}} \delta^{x_2} p(t, T_j^{x_2}) \left(L_j^{x_2}(t) + S \right) \tag{4.34}$$

The fair basis swap spread $S^{BSw}(t; \mathcal{T}^{x_1}, \mathcal{T}^{x_2})$ is thus given by

$$S^{BSw}(t; \mathcal{T}^{x_1}, \mathcal{T}^{x_2}) = \frac{\sum_{i=1}^{M_{x_1}} \delta^{x_1} p(t, T_i^{x_1}) L_i^{x_1}(t) - \sum_{j=1}^{M_{x_2}} \delta^{x_2} p(t, T_j^{x_2}) L_j^{x_2}(t)}{\sum_{j=1}^{M_{x_2}} \delta^{x_2} p(t, T_j^{x_2})} \tag{4.35}$$

Passing now to non-linear derivatives, their pricing is based on the affine property of the driving process under all forward measures and an application of Fourier transform methods for option pricing. The Fourier transform methods for option pricing were discussed already in Chap. 3.

Let us first consider a caplet, as defined in Sect. 1.4.6. A straightforward application of the Fourier transform method yields the following pricing formula, which is proved in Proposition 6.1 in Grbac et al. (2014).

Proposition 4.2 *Consider a tenor x and a caplet with strike K and with payoff* $\delta^x(L_k^x(T_{k-1}^x) - K)^+$ *at time* T_k^x. *Its time-0 price is given by*

$$P^{Cpl}(0; T_k^x, K) = \frac{p(0, T_k^x)}{2\pi} \int_{\mathbb{R}} K_x^{1-\mathscr{R}+iw} \frac{\Theta_{\mathscr{W}_{k-1}^x}(\mathscr{R} - iw)}{(\mathscr{R} - iw)(\mathscr{R} - 1 - iw)} dw \qquad (4.36)$$

for any $\mathscr{R} \in (1, \infty) \cap \mathscr{I}_k^x$, $K_x := 1 + \delta^x K$ *and where*

$$\mathscr{I}_k^x = \left\{ z \in \mathbb{R} : (1 - z)B^{u_k^x}(T_M - T_{k-1}^x) + zB^{v_{k-1}^x}(T_M - T_{k-1}^x) \in \mathscr{I}_T \right\} \qquad (4.37)$$

The random variable $\mathscr{W}_{k-1}^x$ *is defined as*

$$\begin{aligned} \mathscr{W}_{k-1}^x &= \log \left(M_{T_{k-1}^x}^{v_{k-1}^x} / M_{T_{k-1}^x}^{u_k^x} \right) \\ &= A^{v_{k-1}^x}(T_M - T_{k-1}^x) - A^{u_k^x}(T_M - T_{k-1}^x) \\ &\quad + \left\langle B^{v_{k-1}^x}(T_M - T_{k-1}^x) - B^{u_k^x}(T_M - T_{k-1}^x), X_{T_{k-1}^x} \right\rangle \\ &=: A + \langle B, X_{T_{k-1}^x} \rangle \end{aligned} \qquad (4.38)$$

with the moment generating function $\Theta_{\mathscr{W}_{k-1}^x}$ *under the measure* $Q^{T_k^x}$ *given by*

$$\Theta_{\mathscr{W}_{k-1}^x}(z) = E^{T_k^x}\left\{ e^{z\mathscr{W}_{k-1}^x} \right\} = E^{T_k^x}\left\{ \exp\left(z(A + \langle B, X_{T_{k-1}^x} \rangle)\right) \right\}$$

which is known explicitly thanks to the affine property of the model.

Note that the pricing formula (4.36) has an arbitrary $\mathscr{R} \in (1, \infty) \cap \mathscr{I}_k^x$ on the right-hand side. Theoretically, the value of the right-hand side does not depend on the specific choice of $\mathscr{R}$. However, different choices of $\mathscr{R}$ may affect the efficiency of the numerical implementation.

Regarding swaption pricing in the affine multiple curve Libor model, let us consider a swaption as defined in Sect. 1.4.7. The time-0 price of a swaption with exercise date T_0^x and swap rate R, written on an underlying swap with tenor structure $\mathscr{T}^x$, is given by

$$P^{Swn}(0; T_0^x, \mathscr{T}^x, R) = p(0, T_0^x) E^{Q_0^{T_0^x}}\left\{ \left(P^{Sw}(T_0^x; \mathscr{T}^x, R)\right)^+ \right\}$$

where $P^{Sw}(T_0^x; \mathscr{T}^x, R)$ is the price of the underlying swap at time T_0^x which can be expressed as follows

$$P^{Sw}(T_0^x; \mathscr{T}^x, R) = \delta^x \sum_{k=1}^{M_x} p(T_0^x, T_k^x)(L_k^x(T_0^x) - R) = \sum_{k=1}^{M_x} \frac{M_{T_0^x}^{v_{k-1}^x}}{M_{T_0^x}^{u_0^x}} - \sum_{k=1}^{M_x} R_x \frac{M_{T_0^x}^{u_k^x}}{M_{T_0^x}^{u_0^x}}$$

where $R_x := 1 + \delta^x R$. Here we have used (4.20), (4.22) and the telescopic product to obtain

$$p(T_0^x, T_k^x) = \frac{p(T_0^x, T_k^x)}{p(T_0^x, T_{k-1}^x)} \cdots \frac{p(T_0^x, T_1^x)}{p(T_0^x, T_0^x)} = \frac{M_{T_0^x}^{u_k^x}}{M_{T_0^x}^{u_0^x}} \qquad (4.39)$$

and Eq. (4.23) for $L_k^x(T_0^x)$. Therefore, for the swaption price we have

$$P^{Swn}(0; T_0^x, \mathscr{T}^x, R) = p(0, T_0^x) \, E^{Q^{T_0^x}} \left\{ \left(\sum_{k=1}^{M_x} \frac{M_{T_0^x}^{v_{k-1}^x}}{M_{T_0^x}^{u_0^x}} - \sum_{k=1}^{M_x} R_x \frac{M_{T_0^x}^{u_k^x}}{M_{T_0^x}^{u_0^x}} \right)^+ \right\}$$

$$= p(0, T_M) \, E^{Q^{T_M}} \left\{ \left(\sum_{k=1}^{M_x} M_{T_0^x}^{v_{k-1}^x} - \sum_{k=1}^{M_x} R_x M_{T_0^x}^{u_k^x} \right)^+ \right\} \qquad (4.40)$$

where the second equality follows by a measure change from $Q^{T_0^x}$ to Q^{T_M}, cf. (4.17).

Evaluating the above expectation is a computationally demanding task, due to the high-dimensionality of the problem. However, in order to arrive at semi-closed pricing formulas based on the affine property of the model and the Fourier transform methods, an efficient and accurate linear boundary approximation developed in Singleton and Umantsev (2002) can be used. Numerical results for this approximation are reported in Grbac et al. (2014) and below we describe the method and cite the main result.

Firstly, one defines the probability measures $\overline{Q}^{T_k^x}$, for every $k \in \mathscr{K}^x$, by the Radon–Nikodym density

$$\left. \frac{d\overline{Q}^{T_k^x}}{dQ^{T_M}} \right|_{\mathscr{F}_t} = \frac{M_t^{v_k^x}}{M_0^{v_k^x}} \qquad (4.41)$$

The process X, defined by its components in (4.13), is a time-inhomogeneous affine process under every $\overline{Q}^{T_k^x}$, which can be shown exactly in the same way as for the forward measures $Q^{T_k^x}$. The expectation with respect to the measure $\overline{Q}^{T_k^x}$ will be denoted by $\overline{E}^{T_k^x}$ below.

Next, starting from the second equality in (4.40) and using the definitions of martingales $M^{u_k^x}$ and $M^{v_{k-1}^x}$ given in (4.18) and (4.19), one defines the function $f :$ $\mathbb{R}_+^d \to \mathbb{R}$ by

$$f(y) = \sum_{i=1}^{M_x} \exp\left(A^{v_{i-1}^x}(T_M - T_0^x) + \langle B^{v_{i-1}^x}(T_M - T_0^x), y \rangle\right)$$

$$- \sum_{i=1}^{M_x} R_x \exp\left(A^{u_i^x}(T_M - T_0^x) + \langle B^{u_i^x}(T_M - T_0^x), y \rangle\right) \tag{4.42}$$

This function determines the exercise boundary for the price of the swaption. Since the characteristic function of $f(X_{T_0^x})$ cannot be computed explicitly, the method of Singleton and Umantsev (2002) is used and f is approximated by a linear function. More precisely, one has

$$f(X_{T_0^x}) \approx \widetilde{f}(X_{T_0^x}) := \mathcal{C} + \langle \mathcal{D}, X_{T_0^x} \rangle \tag{4.43}$$

where the constants $\mathcal{C}$ and $\mathcal{D}$ are determined according to the linear regression procedure described in Singleton and Umantsev (2002, pp. 432–434). The line $\langle \mathcal{D}, X_{T_0^x} \rangle = -\mathcal{C}$ approximates the exercise boundary, hence $\mathcal{C}, \mathcal{D}$ are strike-dependent. Let $\Im(z)$ denote the imaginary part of a complex number $z \in \mathbb{C}$. Now, we have the following result.

Proposition 4.3 *Assume that* $\mathcal{C}$, $\mathcal{D}$ *are determined by the approximation (4.43). The price of the swaption with swap rate R, option maturity* T_0^x, *on a swap with tenor structure* $\mathcal{T}^x$, *is approximated by*

$$P^{Swn}(0; T_0^x, \mathcal{T}^x, K) \approx p(0, T_M) \sum_{i=1}^{M_x} M_0^{v_{i-1}^x} \left[\frac{1}{2} + \frac{1}{\pi} \int_0^\infty \frac{\Im\left(\widetilde{\xi}_{i-1}^x(z)\right)}{z} dz \right]$$

$$- R_x \sum_{i=1}^{M_x} p(0, T_i^x) \left[\frac{1}{2} + \frac{1}{\pi} \int_0^\infty \frac{\Im\left(\widetilde{\zeta}_i^x(z)\right)}{z} dz \right] \tag{4.44}$$

where $\widetilde{\zeta}_i^x$ *and* $\widetilde{\xi}_i^x$ *approximate the characteristic functions*

$$\zeta_i^x(z) := E^{T_i^x}\left\{\exp\left(izf(X_{T_0^x})\right)\right\} \quad and \quad \xi_i^x(z) := \overline{E}^{T_i^x}\left\{\exp\left(izf(X_{T_0^x})\right)\right\}$$

and are given by

$$\widetilde{\zeta}_i^x(z) := E^{T_i^x}\left\{\exp\left(iz\widetilde{f}(X_{T_0^x})\right)\right\}$$
$$= \exp\left(iz\mathcal{C} + A^{B^{u_i^x}(T_M - T_0^x) + iz\mathcal{D}}(T_0^x) - A^{B^{u_i^x}(T_M - T_0^x)}(T_0^x)\right.$$
$$\left. + \langle B^{B^{u_i^x}(T_M - T_0^x) + iz\mathcal{D}}(T_0^x) - B^{B^{u_i^x}(T_M - T_0^x)}(T_0^x), X_0 \rangle\right) \tag{4.45}$$

$$\widetilde{\xi}_i^x(z) := \overline{E}^{T_i^x}\left\{\exp\left(iz\widetilde{f}(X_{T_0^x})\right)\right\}$$
$$= \exp\left(iz\mathcal{C} + A^{B^{v_i^x}(T_M - T_0^x) + iz\mathcal{D}}(T_0^x) - A^{B^{v_i^x}(T_M - T_0^x)}(T_0^x)\right.$$
$$\left. + \langle B^{B^{v_i^x}(T_M - T_0^x) + iz\mathcal{D}}(T_0^x) - B^{B^{v_i^x}(T_M - T_0^x)}(T_0^x), X_0 \rangle\right) \tag{4.46}$$

Remark 4.5 Note that an approximate pricing formula for the price of a basis swaption defined in Remark 1.10 can be derived as well. For details we refer to Grbac et al. (2014).

Remark 4.6 (*Calibration*) Another important remark regarding the multiple curve affine model is its flexibility to calibrate to option market data. A specification of the model based on the CIR driving processes with jumps proves to fit very well the caplet data, simultaneously for multiple tenors. Since this issue is beyond the scope of this book, we refer the interested reader to Grbac et al. (2014) for all the details.

4.3 Multiplicative Spread Models

In this section we give an overview of the modeling approaches based on multiplicative spreads. The idea to consider the multiplicative spreads has been first proposed in Henrard (2007, 2010), see also the recent book Henrard (2014). The same choice for the modeling quantities has been made in the recent paper by Cuchiero et al. (2015). Recall from (1.36) of Sect. 1.4.4 (noting that $F(t; T, T + \Delta) = R^{OIS}(t; T, T + \Delta)$) that the multiplicative forward Libor-OIS spreads are defined as

$$\Sigma(t; T, T + \Delta) = \frac{1 + \Delta L(t; T, T + \Delta)}{1 + \Delta F(t; T, T + \Delta)} \qquad (4.47)$$

where as usual

$$L(t; T, T + \Delta) = E^{Q^{T+\Delta}} \{ L(T; T, T + \Delta) \mid \mathscr{F}_t \}$$

denotes the forward Libor rate (FRA rate) and the forward OIS rates are defined via relation (see (1.16))

$$1 + \Delta F(t; T, T + \Delta) = \frac{p(t, T)}{p(t, T + \Delta)}$$

Note that the notation $\Sigma(t; T, T + \Delta)$ in (4.47) corresponds to the notation $S^{\Delta}(t, T)$ in Cuchiero et al. (2015). As Henrard (2014) and Cuchiero et al. (2015) point out, the choice of multiplicative spreads as modeling quantities instead of the forward Libor rates is made for the convenience of modeling. Empirical findings on the positivity and monotonicity of the additive spreads with respect to the tenor Δ motivate one to model directly the spreads instead of the forward Libor rates in order to access more easily those two features, cf. also the comments in Sect. 4.1. Passing from the additive to the multiplicative spreads still serves the same purpose, while allowing for more analytical tractability in the model. Moreover, as noticed by Cuchiero et al. (2015), the multiplicative spreads are related to the forward exchange rates when the multiple curve market is considered in a foreign exchange analogy (see Sect. 3.3.1.2, Remark 3.8).

In the work of Henrard, expressions for the prices of various interest rate derivatives in terms of the multiplicative spreads have been developed. For the dynamics of the spreads, Henrard introduces some assumptions that, from a modeling point of view, appear to be rather restrictive. The assumptions are, in particular: the spreads are supposed to be independent from the forward OIS rates $F(t; T, T + \Delta)$, and for tractable pricing of optional derivatives, an assumption of the spreads being constant for each maturity is introduced in addition. As stated in Sect. 7.3 of Henrard (2014), this has the advantage of allowing to determine the price of any instrument in the post-crisis setting by directly applying the corresponding pre-crisis formula (in the case of optional derivatives one has only to scale the strike). To model the OIS bond price dynamics Henrard (2010) considers an HJM 1-factor Gaussian framework; cf. Sect. 2 therein.

The framework proposed in Cuchiero et al. (2015), that we shall describe in more detail below, allows for more modeling flexibility and, in fact, it can be shown that many of the existing modeling approaches can be recovered from their setting. To this effect the authors develop a general semimartingale HJM framework for the multiple curve term structure, which is inspired by Kallsen and Krühner (2013). Their approach is situated in between the HJM and the LMM approaches and in this sense is similar to the approach taken in Sect. 3.2. Concerning the model choice for the dynamics of the spreads, the affine specification of the framework by Cuchiero et al. (2015) can also be seen as a possible extension to continuous tenors of the model from Sect. 4.2 (see Remark 4.8).

Let us now give an overview of the framework proposed by Cuchiero et al. (2015). The modeling quantities are the OIS bonds $p(t, T)$ and the multiplicative spreads $\Sigma(t; T, T + \Delta)$. They consider a finite number of tenors denoted by $\Delta_1, \ldots, \Delta_m$ (corresponding to tenors from 1 day to 12 months). The framework allows to reproduce main features of the multiplicative spreads observed in the market (see (4.49)):

$$\Sigma(t; T, T + \Delta_i) \geq 1 \quad \text{and} \quad \Sigma(t; T, T + \Delta_i) \geq \Sigma(t; T, T + \Delta_j), \quad \text{for } \Delta_i \geq \Delta_j$$

The first property is equivalent to the positivity of additive spreads and the second one is the monotonicity with respect to the tenor. Moreover, the definition (4.47) of the spread $\Sigma(t; T, T + \Delta_i)$ implies that it has to be a Q^T-martingale because the forward Libor rate $1 + \Delta L(\cdot; T, T + \Delta)$ is a $Q^{T+\Delta}$-martingale and $\frac{dQ^T}{dQ^{T+\Delta}}\big|_{\mathscr{F}_t} = \frac{1+\Delta F(t;T,T+\Delta)}{1+\Delta F(0;T,T+\Delta)}$ by (1.15) and (1.16).

As stated above, to develop their framework, Cuchiero et al. (2015) make use of the classical HJM setup presented in the philosophy of Kallsen and Krühner (2013). The main idea behind this approach is to identify "canonical" assets which are the underlyings for the assets of interest and then obtain a convenient parametrization (a "codebook" as referred to in Kallsen and Krühner 2013) of the related term structures. In order to do so, one first specifies simple elementary models for the term structure of the canonical assets to understand the general relations that have to hold between the fundamental modeling quantities and thus obtains the codebooks; then one prescribes a stochastic evolution for the codebooks, which has to satisfy certain consistency conditions. Let us illustrate this approach on the first fundamental quantities in the framework of Cuchiero et al. (2015), which are the OIS bonds $p(t, T)$ (a similar procedure is then repeated for the second fundamental modeling quantities, i.e. the multiplicative spreads). The underlying canonical asset for the OIS bonds is the OIS short rate r. The idea is thus to exploit the connection of the OIS bond prices and the OIS bank account $B_t = \exp(\int_0^t r_s ds)$, supposing firstly that r is a deterministic short rate. This yields the relation $r_T = -\frac{\partial}{\partial T} \log(p(t, T))$, which is the codebook for the bond prices. Now, since market data indicate that $-\frac{\partial}{\partial T} \log(p(t, T))$ evolves randomly over time, this leads to instantaneous forward rates $f_t(T) = -\frac{\partial}{\partial T} \log(p(t, T))$, for which then a stochastic model is specified. Setting for the instantaneous forward rate $f_t(T) = -\frac{\partial}{\partial T} \log(p(t, T)) = -\eta_t(T)$ and $Z_t = -\log B_t = -\int_0^t r_s ds$ for the short rate r, Cuchiero et al. (2015) specify an HJM OIS bond price model given by a quintuple $(Z, \eta_0, \alpha, \sigma, X)$ such that

$$\frac{p(t, T)}{B_t} = e^{Z_t + \int_t^T \eta_t(u) du}$$

with

$$\eta_t(T) = \eta_0(T) + \int_0^t \alpha_s(T) ds + \int_0^t \sigma_s(T) dX_s \qquad (4.48)$$

where (X, Z) is a general multidimensional semimartingale with absolutely continuous characteristics (Itô semimartingale) and the processes α and σ satisfy the implicit measurability and integrability conditions, together with a suitable HJM drift condition to ensure absence of arbitrage in the OIS bonds. This in particular yields

$$\frac{p(t, T)}{B_t} = E\left\{ e^{Z_T} \mid \mathscr{F}_t \right\}$$

Similarly, for the multiplicative spread $\Sigma(t; T, T + \Delta_i)$, passing via a suitable codebook, Cuchiero et al. (2015) again obtain an HJM-type model given by a quintuple $(Z^i, \eta_0^i, \alpha^i, \sigma^i, X)$ such that

$$\Sigma(t; T, T + \Delta_i) = e^{Z_t^i + \int_t^T \eta_t^i(u) du}$$

with η^i having a dynamics similar to (4.48) with corresponding α^i and σ^i, which satisfy a drift condition ensuring the Q^T-martingale property of $\Sigma(\cdot; T, T + \Delta_i)$. The quantity $Z_t^i = \log(\Sigma(t; t, t + \Delta_i))$ can be seen as the log-spot spread and $-\eta_t^i(T) = -\frac{\partial}{\partial T} \log(\Sigma(t; T, T + \Delta_i))$ as the forward spread rate, by analogy to the OIS bond price model. The martingale property of $\Sigma(\cdot; T, T + \Delta_i)$ yields

$$\Sigma(t; T, T + \Delta_i) = E^{Q^T}\left\{ e^{Z_T^i} \mid \mathscr{F}_t \right\}$$

In order to specify the model further, the semimartingales Z^i are assumed to be of the following form

$$Z_t^i = e^{\langle u_i, Y_t \rangle}$$

with a common n-dimensional Itô semimartingale Y for all i, and $u_1, \ldots, u_m$ given vectors in $\mathbb{R}^n$. A common driving process Y for all tenors Δ_i is a choice which allows to capture the interdependencies between the spreads associated to different tenors. The vectors u_i enable one to implement easily the ordered spreads $1 \leq \Sigma(t; T, T + \Delta_1) \leq \cdots \leq \Sigma(t; T, T + \Delta_m)$. More precisely, one would have to consider a process Y taking values in some cone $C \subset \mathbb{R}^n$ and vectors $u_i \in C^*$ such that $0 \prec u_1 \prec \cdots \prec u_m$, where C^* denotes the dual cone of C with the order relation $\prec$. This then easily implies

$$1 \leq \Sigma(t; T, T + \Delta_i) = E^{Q^T} \left\{ e^{\langle u_i, Y_T \rangle} \middle| \mathscr{F}_t \right\} \leq E^{Q^T} \left\{ e^{\langle u_j, Y_T \rangle} \middle| \mathscr{F}_t \right\} = \Sigma(t; T, T + \Delta_j) \tag{4.49}$$

for $\Delta_i < \Delta_j$.

Remark 4.7 Making use of relation (4.47), the payoffs of all linear and optional interest rate derivatives can be expressed as functions of the OIS bond prices $p(t, T)$ and the multiplicative spreads $\Sigma(t; T, T + \Delta_i)$, cf. Cuchiero et al. (2015) for explicit expressions in the general framework. The prices of linear derivatives can thus easily be expressed in terms of these modeling quantities.

Regarding optional derivatives, to obtain a tractable specification of the general framework, Cuchiero et al. (2015) suggest the class of affine processes as driving processes, which allows convenient pricing by standard techniques resorting to the Fourier transform.

Remark 4.8 Even though the approach proposed by Cuchiero et al. (2015) can be situated in the HJM framework, its affine specification can also be regarded as a continuous tenor extension of the affine Libor model from Sect. 4.2 with the difference that the quantities modeled here are not the forward Libor rates, but rather the multiplicative spreads which in this case are given by

$$\Sigma(t; T, T + \Delta_i) = \frac{M_t^{v(T, \Delta_i)}}{M_t^{u(T)}}$$

where $u(\cdot)$ and $v(\cdot, \Delta_i)$ are mappings from $[0, T]$ to $\mathbb{R}^d$. Similarly to Sect. 4.2, imposing conditions on these mappings allows to ensure positivity and monotonicity of the spreads in the model.

References

J. Akahori, Y. Hishida, J. Teichmann, T. Tsuchiya, A heat kernel approach to interest rate models. Japan J. Ind. Appl. Math. **2**, 419–439 (2014)

F.M. Ametrano, M. Bianchetti, Everything you always wanted to know about multiple interest rate curve bootstrapping but were afraid to ask. Preprint, SSRN/2219548 (2013)

M. Bianchetti, Two curves, one price, in *Risk Magazine* (2010) pp. 74–80

M. Bianchetti, M. Morini (eds.), *Interest Rate Modelling After the Financial Crisis* (Risk Books, London, 2013)

T.R. Bielecki, M. Rutkowski, *Credit Risk: Modeling, Valuation and Hedging* (Springer, Berlin, 2001)

T. Björk, *Arbitrage Theory in Continuous Time*, 3rd edn. (Oxford University Press, Oxford, 2009)

T. Björk, Y. Kabanov, W. Runggaldier, Bond market structure in the presence of marked point processes. Math. Financ. **7**, 211–239 (1997)

A. Brace, D. Gątarek, M. Musiela, The market model of interest rate dynamics. Math. Financ. **7**, 127–155 (1997)

D. Brigo, F. Mercurio, *Interest Rate Models—Theory and Practice*, 2nd edn. (Springer, Berlin, 2006)

D. Brigo, M. Morini, A. Pallavicini, *Counterparty Credit Risk, Collateral and Funding: With Pricing Cases For All Asset Classes* (Wiley, New York, 2013)

D.C. Brody, L.P. Hughston, Chaos and coherence: a new framework for interest-rate modelling. Proc. R. Soc. A **460**, 85–110 (2004)

A.J.G. Cairns, *Interest Rate Models: An Introduction* (Princeton University Press, Princeton, 2004)

P. Carr, D.B. Madan, Option valuation using the fast Fourier transform. J. Comput. Financ. **2**(4), 61–73 (1999)

Y. Chang, E. Schlögl, A consistent framework for modelling basis spreads in tenor swaps. Preprint, SSRN/2433829 (2015)

L. Chen, D. Filipović, H.V. Poor, Quadratic term structure models for risk-free and defaultable rates. Math. Financ. **14**, 515–536 (2004)

R. Cont, R. Mondescu, Y. Yu, Central clearing of interest rate swaps: a comparison of offerings. Preprint, SSRN/1783798 (2011)

J.C. Cox, J.E. Ingersoll, S.A. Ross, A theory of the term structure of interest rates. Econometrica **53**, 385–407 (1985)

S. Crépey, Bilateral counterparty risk under funding constraints—Part II: CVA. Math. Financ. **25**(1), 23–50 (2015)

© The Author(s) 2015

Z. Grbac and W.J. Runggaldier, *Interest Rate Modeling: Post-Crisis Challenges and Approaches*, SpringerBriefs in Quantitative Finance, DOI 10.1007/978-3-319-25385-5

S. Crépey, R. Douady, LOIS: credit and liquidity, in *Risk Magazine* (2013) pp. 82–86

S. Crépey, Z. Grbac, H.N. Nguyen, A multiple-curve HJM model of interbank risk. Math. Financ. Econ. **6**(3), 155–190 (2012)

S. Crépey, T.R. Bielecki, D. Brigo, *Counterparty Risk and Funding—A Tale of Two Puzzles* (Chapman & Hall/CRC Financial Mathematics Series, Boca Raton, 2014)

S. Crépey, Z. Grbac, N. Ngor, D. Skovmand, A Lévy HJM multiple-curve model with application to CVA computation. Quant. Financ. **15**(3), 401–419 (2015a)

S. Crépey, A. Macrina, T.M. Nguyen, D. Skovmand, Rational multi-curve models with counterparty-risk valuation adjustments. Forthcoming in *Quantitative Finance* (2015b). Preprint available at arXiv:1502.07397

C. Cuchiero, C. Fontana, A. Gnoatto, A general HJM framework for multiple yield curve modeling, in *Finance and Stochastics*, (2015, Forthcoming)

E. Eberlein, K. Glau, A. Papapantoleon, Analysis of Fourier transform valuation formulas and applications. Appl. Math. Financ. **17**(3), 211–240 (2010)

N. El Karoui, R. Myneni, R. Viswanathan, Arbitrage pricing and hedging of interest rate claims with state variables, theory and applications. Working Paper (1992)

D. Filipović, *Term-Structure Models* (Springer, Berlin, 2009)

D. Filipović, A.B. Trolle, The term structure of interbank risk. J. Financ. Econ. **109**(3), 707–733 (2013)

D. Filipović, M. Larsson, A.B. Trolle, Linear-rational term structure models. Preprint, SSRN/2397898 (2014)

B. Flesaker, L.P. Hughston, Positive interest. Risk **9**(1), 46–49 (1996)

M. Fujii, Y. Shimada, A. Takahashi, A note on construction of multiple swap curves with and without collateral. FSA Res. Rev. **6**, 139–157 (2010)

M. Fujii, Y. Shimada, A. Takahashi, A market model of interest rates with dynamic basis spreads in the presence of collateral and multiple currencies. Wilmott Mag. **54**, 61–73 (2011)

J. Gallitschke, S. Müller, F.T. Seifried, Post-crisis interest rates: XIBOR mechanics and basis spreads. Preprint, SSRN/2448657 (2014)

R.M. Gaspar, General quadratic term structures for bond, futures and forward prices. 2004. SSE/EFI Working paper Series in Economics and Finance 559

H. Geman, N. El Karoui, J.-C. Rochet, Changes of numéraire, changes of probability measures and option pricing. J. Appl. Probab. **32**, 443–458 (1995)

A. Gombani, W.J. Runggaldier, A filtering approach to pricing in multifactor term structure models. Int. J. Theor. Appl. Financ. **4**, 303–320 (2001)

M. Grasselli, G. Miglietta, A flexible spot multiple-curve model. Forthcoming in *Quantitative Finance* (2015)

Z. Grbac, A. Papapantoleon, D. Skovmand, J. Schoenmakers, Affine LIBOR models with multiple curves: theory, examples and calibration. SIAM J. Financ. Math. **6**, 984–1025 (2015)

Z. Grbac, L. Meneghello, W.J. Runggaldier, Derivative pricing for a multicurve extension of the Gaussian, exponentially quadratic short rate model. Forthcoming in the Conference Proceedings volume *Challenges in Derivatives Markets – Fixed income modeling, valuation adjustments, risk management, and regulation* (Springer, 2016). Preprint available at arXiv:1512.03259

D. Heath, R. Jarrow, A. Morton, Bond pricing and the term structure of interest rates: a new methodology for contingent claims valuation. Econometrica **60**, 77–105 (1992)

D. Heller, N. Vause, Expansion of central clearing, in *BIS Quarterly Review* (2011)

M. Henrard, The irony in the derivatives discounting. Wilmott Mag. **30**, 92–98 (2007)

M. Henrard, The irony in the derivatives discounting part II: the crisis. Wilmott Mag. **2**, 301–316 (2010)

M. Henrard, *Interest Rate Modelling in the Multi-Curve Framework: Foundations, Evolution and Implementation* (Palgrave Macmillan, London, 2014)

L.P. Hughston, A. Macrina, Pricing fixed-income securities in an information-based framework. Appl. Math. Financ. **19**(4), 361–379 (2012)

J. Hull, A. White, Pricing interest-rate-derivative securities. Rev. Financ. Stud. **3**(4), 573–592 (1990)

J. Hull, A. White, Libor vs OIS: the derivatives discounting dilemma. Preprint, SSRN/2333754 (2013)

P.E. Hunt, J.E. Kennedy, A. Pelsser, Markov-functional interest rate models. Financ. Stoch. **4**, 391–408 (2000)

P.J. Hunt, J.E. Kennedy, *Financial Derivatives in Theory and Practice*, 2nd edn. (Wiley, Chichester, 2004)

J. Kallsen, P. Krühner, On a Heath-Jarrow-Morton approach for stock options. Financ. Stochast. **19**(3), 583–615 (2015)

M. Keller-Ressel, A. Papapantoleon, J. Teichmann, The affine LIBOR models. Math. Financ. **23**, 627–658 (2013)

C. Kenyon, Short-rate pricing after the liquidity and credit shocks: including the basis, in *Risk Magazine* (2010) pp. 83–87

M. Kijima, Y. Muromachi, Reformulation of the arbitrage-free pricing method under the multi-curve environment. Preprint (2015)

M. Kijima, K. Tanaka, T. Wong, A multi-quality model of interest rates. Quant. Financ. **9**(2), 133–145 (2009)

D. Lamberton, B. Lapeyre, *Introduction to Stochastic Calculus Applied to Finance* (Chapman and Hall/CRC, Boca Raton, 2007)

M. Leippold, L. Wu, Asset pricing under the quadratic class. J. Financ. Quant. Anal. **37**(2), 271–291 (2002)

A. McNeil, R. Frey, P. Embrechts, *Quantitative Risk Management: Concepts, Techniques and Tools* (Princeton University Press, Princeton, 2005)

F. Mercurio, Interest rates and the credit crunch: new formulas and market models. Preprint, SSRN/1332205 (2009)

F. Mercurio, A LIBOR market model with stochastic basis, in *Risk Magazine* (2010a) pp. 84–89

F. Mercurio, LIBOR market models with stochastic basis. Preprint, SSRN/1563685 (2010b)

F. Mercurio, Modern LIBOR market models: using different curves for projecting rates and for discounting. Int. J. Theor. Appl. Financ. **13**, 113–137 (2010c)

F. Mercurio, Z. Xie, The basis goes stochastic, in *Risk* (2012) pp. 78–83

F.-L. Michaud, C. Upper, What drives interbank rates? Evidence from the Libor panel, in *BIS Quarterly Review* (2008) pp. 47–58

G. Miglietta, Topics in Interest Rate Modeling. Ph.D. thesis, University of Padova (2015)

K.R. Miltersen, K. Sandmann, D. Sondermann, Closed form solutions for term structure derivatives with log-normal interest rates. J. Financ. **52**, 409–430 (1997)

N. Moreni, A. Pallavicini, Parsimonious HJM modelling for multiple yield-curve dynamics. Quant. Financ. **14**(2), 199–210 (2014)

M. Morini, Solving the puzzle in the interest rate market. Preprint, SSRN/1506046 (2009)

L. Morino, W.J. Runggaldier, On multicurve models for the term structure, in *Nonlinear Economic Dynamics and Financial Modelling*, ed. by R. Dieci, X.Z. He, C. Hommes (Springer, Berlin, 2014), pp. 275–290

M. Musiela, M. Rutkowski, *Martingale Methods in Financial Modelling*, 2nd edn. (Springer, Berlin, 2005)

T.A. Nguyen, F. Seifried, The multi-curve potential model. Preprint, SSRN/2502374 (2015)

A. Pallavicini, D. Brigo, Interest-rate modelling in collateralized markets: multiple curves, credit-liquidity effects, CCPs. Preprint (2013) arXiv:1304.1397

A. Pallavicini, M. Tarenghi, Interest-rate modeling with multiple yield curves. Preprint (2010) arXiv:1006.4767

A. Papapantoleon, Old and new approaches to LIBOR modeling. Stat. Neerl. **64**, 257–275 (2010)

A. Pelsser, A tractable yield-curve model that guarantees positive interest rates. Rev. Deriv. Res. **1**, 269–284 (1997)

V. Piterbarg, Funding beyond discounting: collateral agreements and derivatives pricing. Risk Mag. **2**, 97–102 (2010)

E. Platen, D. Heath, *A Benchmark Approach to Quantitative Finance* (Springer, Berlin, 2010)

E. Platen, S. Tappe, Real-world forward rate dynamics with affine realizations, in *Stochastic Analysis and Applications* (2015, Forthcoming)

L. Chris, G. Rogers, The potential approach to the term structure of interest rates and foreign exchange rates. Math. Financ. **7**, 157–176 (1997)

K. Singleton, L. Umantsev, Pricing coupon-bond options and swaptions in affine term structure models. Math. Financ. **12**, 427–446 (2002)

O. Vasiček, An equilibrium characterization of the term structure. J. Financ. Econ. **5**(2), 177–188 (1977)

Printed in the United States
By Bookmasters